The Polyvagal Theory

The Simplified Guide to Understanding the Autonomic Nervous System and the Healing Power of the Vagus Nerve - Learn to Manage Emotional Stress and PTSD Through Neurobiology

© **Copyright 2019 - All rights reserved.**

The content contained within this book may not be reproduced, duplicated or transmitted without direct written permission from the author or the publisher.

Under no circumstances will any blame or legal responsibility be held against the publisher, or author, for any damages, reparation, or monetary loss due to the information contained within this book, either directly or indirectly.

Legal Notice:
This book is copyright protected. It is only for personal use. You cannot amend, distribute, sell, use, quote or paraphrase any part, or the content within this book, without the consent of the author or publisher.

Disclaimer Notice:
Please note the information contained within this document is for educational and entertainment purposes only. All effort has been executed to present accurate, up to date, reliable, complete information. No warranties of any kind are declared or implied. Readers acknowledge that the author is not engaging in the rendering of legal, financial, medical or professional advice. The content within this book has been derived from various sources. Please consult a licensed professional before attempting any techniques outlined in this book.

By reading this document, the reader agrees that under no circumstances is the author responsible for any losses, direct or indirect, that are incurred as a result of the use of the information contained within this document, including, but not limited to, errors, omissions, or inaccuracies.

Table of Contents

Introduction 1
   What is the Polyvagal Theory? 3
   Why Does it Matter to You? 5
   How Can It Be Applied? 6

Chapter 1: The Autonomic Nervous System 10
   The Central and Peripheral Nervous Systems 11
   The Sympathetic Nervous System 14
   The Parasympathetic Nervous System 17
   Ventral and Dorsal Vagal Complexes 19

Chapter 2: Communication and Connection 24
   Social Communication Systems 26
   Social Engagement Systems 30
   Polyvagal Interactions 38

Chapter 3: The Physiological Regulation of Emotion 43
   Managing Stress 44
   Overcoming Depression 51
   Asperger's Spectrum and Autism 54

Chapter 4: Self-Regulation Vagal Applications .. 58
   Yoga Poses and Stretches as Therapy 59
   Meditative Vagal Stimulation 64
   Auricular Acupuncture 69
   Diaphragmatic Exercises 72
   Carotid Sinus Massage 74
   Vocal Cord and Facial Stimulation 76

Chapter 5: How Trauma Affects the Nervous System 78
   Nervous System Overload 79
   Degrees Of Stress 84
   Panic And Hyperactivity 91

Chapter 6: The Polyvagal Theory And PTSD 94
   The Three-Part Brain 96
   Post-Traumatic Brain Reeducation 100
   The Parasympathetic Recovery 103
   Reading Body Language 106

Chapter 7: The Polyvagal Theory And Emotional Stress 110
   Normal or Interactive State 110
   Emotional Stress-Induced Sympathetic Response 115
   Shutdown or Calming Parasympathetic Response 117

Chapter 8: The Healing Power of Vagal Tone .. 124
   Regulating Emotion 125
   Cardiovascular Applications 129
   Autoimmune Responses and Inflammation 133

Chapter 9: Clinical Applications of Polyvagal Theory 137
   Facial Expressions, Asperger's Spectrum, and Autism 137
   Vagus Nerve Dietary and Nutritional Influences ... 139
   Electrical Vagus Nerve Stimulation 141

Conclusion 145
   Polyvagal Applications You Can Practice 146
   Medical Applications 148
   Non-Medical Applications 150

Reference List 154

Introduction

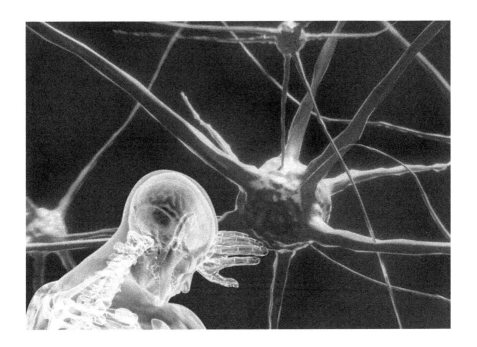

In 1994, Dr. Stephen Porges, who was the director of the Brain-Body Center at Illinois University in Chicago, developed a unique perspective on the autonomic nervous system (ANS). Until that point, the ANS was thought to be comprised of two systems or response mechanisms: the sympathetic, action-initiating, and the parasympathetic, deactivating and calming nervous systems. Porges determined there is a third, extreme ANS response, which freezes and immobilizes the individual. He also determined that the vagus nerve, which is the 10th and longest, most diverse of the 12 cranial nerves emanating primarily from the brainstem, mediates or influences two of the three systems. One is the parasympathetic nervous system, which deactivates the action and energy of the sympathetic response, replacing it with a calming response system. This relaxing effect is mediated by a frontal branch of the vagus nerve, which Porges named the ventral vagal.

The rear branch of the vagus nerve, the dorsal vagal, is responsible for the primitive, shutting down and freezing responses to threats and extreme emotional and physical stress.

Porge's discoveries have the potential for both involuntary and voluntary actions that can affect our physical and mental states. Applications range from simple relaxation and calming techniques (that you can easily learn and put into daily practice) to professional medical treatment of diseases and disorders, among these are anxiety, depression, and Asperger's Spectrum, including autism. Polyvagal Theory applications include facial expressions and body language, as well as manual stimulation of the vagus nerve to achieve vagal tone. These can induce emotional responses, leading to the emergence of social engagement possibilities not previously attainable among autistic patients, especially children.

What is the Polyvagal Theory?

From an evolutionary standpoint, the sympathetic nervous system developed first as the body's urgent response to danger, frequently referred to as the *fight or flight* response. It is automatically and involuntarily activated when an imminent life and death, or otherwise serious threat, is perceived. Fifty thousand years ago, when our Homo Sapiens ancestors were confronted with a wild animal or an aggressive, potentially violent adversary, the sympathetic nervous system stimulated a rapidly elevated heart rate and breathing rate that increased blood pressure, while simultaneously suppressing the metabolism. These actions effectively shut down key visceral activities, including digestion, to divert energy to the more immediate situation. Hormones, including adrenaline, were released to increase energy stores for action further.

While our species today does not regularly confront life-threatening situations, our sympathetic responses remain ready to activate when stimulated by stress, frustration, anger, panic and anxiety. The same involuntary fight or flight mechanisms come into play and we can still get pumped up in any number of situations and confrontations.

More recently in our species' evolutionary history, the parasympathetic nervous system evolved as a counter to the sympathetic response, inducing relaxation by bringing down heart rate and breathing rate, lowering elevated blood pressure, reactivating the digestive system, and creating a sense of calm. Preparedness of fight or flight is replaced by what is called *rest and digest*. The body attains a state of homeostasis, or normality. The parasympathetic response is mediated by the ventral vagus nerve, which has branches reaching the heart, lungs, diaphragm and the entire digestive system.

Dr. Porges' work in developing the Polyvagal Theory led him to discover a third ANS response, called the freeze, mediated by another branch of the vagus nerve, called the dorsal, a nearly total shutting down of the individual's ability to react and respond. Interestingly, it may appear to be an overreaction of the sympathetic response's call to action, but Porges positions it as an extreme parasympathetic response of calming the body too much. The individual may enter a state of panic that prevents movement, speech, any action at all, freezing in place, immobilization, or fainting. Involuntary urination may occur, and the person might even go into shock, which, if severe and prolonged, may lead to cardiac arrest and death.

Among other mammals and reptiles facing a situation with no apparent escape, the response may be to freeze in place and play dead until the threat disappears, or an escape route becomes apparent. This playing dead act may be involuntary, being activated before the creature is even conscious of the situation.

Why Does it Matter to You?

One of the key breakthroughs of the Polyvagal Theory is the discovery that sympathetic responses may be brought under control voluntarily. Conditions ranging from anger, frustration, tension, and aggravation can be eased into a state of calm though mental and physical exercises. These exercises may be practiced alone, at home or while traveling, even at work if a quiet location can be found. Some find that stepping up the level of exercise is beneficial—cardiovascular training that can range from daily walks to more strenuous forms of working out.

If your concerns for your well-being extend beyond simply trying to stay calm and positive in the moment or if you are affected by extreme stress and anxiety, depression and more serious disorders, Polyvagal Theory applications may extend to professional massages and manipulation of the vagus nerves. Autistic patients may respond positively to facial expression and body language approaches that put them at ease, and enable social engagement not previously achieved through other methods and medications.

How Can It Be Applied?

The Polyvagal Theory has raised the level of credibility of a variety of physical and mental practices, bringing a scientific rationale to what was previously thought to be anecdotal, or suggestive results. These methods, which will be presented in detail in subsequent chapters, can be learned and practiced immediately, and in many cases, without the need for equipment or professional training.

One particularly easy yet effective approach is managed breathing, the conscious control of inhales and exhales, the forceful extensions and contractions of the diaphragm, the addition of focused thoughts to drive out intruding negative thoughts. Other techniques include mindfulness, or being in the moment, a state of consciousness of the environment and its sounds, feelings and sensations.

Traditional practices, including Yoga and meditation, which have long been applied to achieve states of peace and calm, are now understood by Polyvagal Theory to apply physical stimuli to the ventral vagus nerve, initiating the parasympathetic response. Both Yoga and meditation emphasize managed breathing, and this practice alone is credited with initiating meaningful physiological and mental changes. Brain scans verify immediate and long-term changes in blood flow in the brain during meditation.

Yoga poses and stretches are also recognized to be encouraging what is called vagal tone, which are the signals to the organs to slow down, relax and achieve a lasting sense of calm. The same positive effects may be achieved by Pilates and other techniques that involve stretching and assuming poses that tense various muscle groups.

Persons whose condition is more serious can benefit from professional assistance. Chiropractic doctors working with Asperger's patients recognized that Polyvagal Theory is compatible with their salutogenic healthcare model, by supporting the recognition that the body can self-regulate itself and can self-heal under the right conditions. This inspired recognition that the Polyvagal can enable the doctors to tap into the healing potential. They apply neurological exercises that stimulate vagal tone in their patients. This empowers the Asperger's Spectrum patients with new ways to hear, to perceive, to respond to people and situations, to smile, to speak, to maintain eye contact.

On the medical level, electric stimulation of the vagus nerve is successfully being applied to treat epilepsy, notably in cases where the patient has not responded well to drugs. A small electrical device is implanted subcutaneously, and a thin wire extends to connect with the left vagus nerve. When spasm-inducing impulses are detected by the device, it emits an electrical pulse that stimulates the vagus nerve to send a signal to the brainstem, which in turn transmits the impulse to the part of the brain controlling involuntary seizures. To date, results are successful in up to 50% of cases.

Electrical stimulation is also being applied in cases of extreme depression that's not controlled by medication or electroshock therapy. Results in reducing depression have been mixed, and experimentation continues. Other conditions responding in some degree to electrical vagal stimulation include Parkinson's Disease, cluster headaches, rheumatoid arthritis, and irritable bowel disease.

Recently, non-invasive electrical vagus nerve stimulation devices (that do not require surgical implantation) have been developed, promising a wider range of possible applications. European authorities have authorized a range of applications. So far, in the U.S., testing has been mostly limited to treating cluster headaches.

Chapter 1: The Autonomic Nervous System

The autonomic nervous system (ANS) controls what are referred to as visceral internal organs, including the heart, lungs, stomach and intestine, as well as certain muscles, like the diaphragm, which controls breathing rates. It's called autonomic because it functions involuntarily, without conscious control. We do not need to think about our heartbeat or breathing or how food is digested. These and other key metabolic functions happen autonomically. In addition to the major organs, the ANS controls certain movements and functions in the facial regions, including oral expressions, vocal emissions, eye movement, pupil dilation, and the production of saliva and mucus.

Most of the ANS functions engage two essential reactive systems: the sympathetic nervous system, which is a call to action initiator, and the parasympathetic nervous system, now, under Polyvagal Theory, called the ventral vagal response. The ventral vagal stimulates a series of opposing, calming responses, and encourages social engagement and interaction. Both are covered below, and we'll also cover a more recent third response, the dorsal vagal response, called the freeze. It's a total shutdown and immobilization. But first, an overview of the central nervous system will provide a perspective of the components and interconnections that involve the brain, the spinal cord and the billions of neurons that interact within and between them.

The Central and Peripheral Nervous Systems

The nervous system is often referred to as either central or peripheral. The central nervous system is comprised of the brain and the spinal cord, called *central* because it is the overriding controller of all bodily functions. The *peripheral* nervous system is made of the many nerves that lead from the spinal cord and radiate throughout our bodies, extending to every extremity. It influences every organ and muscle, affecting a person's control of breathing, movement, thought and perception, including the senses of sight, hearing, taste, smell and the sensation of touch on our skin.

For clarity, the central nervous system (CNS) is comprised exclusively of the nerves within the brain and spinal cord, while the peripheral nervous system (PNS) is comprised of all the nerves exterior to the brain and spinal cord, reaching every part of the body. Importantly, the two systems are interconnected.

The elemental component of the nervous system is the nerve cell, also called a neuron. It's made up of a cell body containing a nucleus, and a long wire-like axon that extends from the cell body to its eventual termination point somewhere in the body. Many neural connections between an extremity and the central nervous system are made of not one but several neurons. At the end of each axon are small branch-like extensions called dendrites, which emit a chemical signal, a neurotransmitter, to an adjacent neuron's dendrites. In turn, the signal is converted to an electrical signal, which continues through the axon of the next neuron. For example, stepping on a sharp object with a bare foot excites the dendrites of a neuron and an electrical impulse travels up the neuron's long axon toward the spinal cord. At the end of the first neuron, dendrites emit neurotransmitting chemicals that are interpreted by the dendrites of the next neuron in line. The chemical reaction is then converted to an electrical signal to continue the trip to the spinal cord.

The CNS nerves tend to be short, traveling short distances, while PNS nerves are much longer, in the extreme, reaching from the spinal cord to the toes. This greater length is what allows an almost instant response when the dendrites in the toe or sole of the foot announce that you've stepped on something irritating. It can inform you that the object is sharp, causing a specific type of pain, or it's as small an irritant as a grain of dry rice.

The human brain contains about 100 billion neurons, and they combine to create many trillions of neural interactions. These affect every aspect of our physical and mental functions, from heartbeat and laughter to memory and thoughts, from breathing and coughing to dreaming and daydreaming, from the reflex that causes a hand to pull back from a hot surface to the peristaltic contractions that move food through the stomach.

The nervous system also contains non-neural glial cells that give supportive structure to neurons and protect them. Glia cells also repair neurons and maintain the chemical neurotransmitters between neurons. They also produce a material called myelin, which insulates axons, preventing electrical signals from leaking in or leaking out.

Of all the physiological and psychological reactions we experience, the action and calming responses of the sympathetic and parasympathetic nervous systems are the best known and most readily recognized. We'll look at them next.

The Sympathetic Nervous System

Our human species, Home Sapiens, extends back at least 70,000 years and the hominid ancestors who preceded us down the long trail of evolution extend back hundreds of thousands and even millions of years. They—and we—survived because of a range of adaptive behaviors, leading to natural selection of the more resilient, more intelligent, and perhaps, most importantly, the people who were and are most reactive to threats and dangers. These reactions today are commonly referred to as the fight or flight response, suggesting an immediate need for aggression, evasion or escape.

Long ago, those threats were tangible; real dangers of survival against wild animals, fierce human competitors for tribal or territorial rights, and sudden natural emergencies, all of which needed instant, unhesitating responses requiring a maximum output of strength or speed. Being able to jump higher, run faster, swing a heavier club, or other defensive maneuvers could ensure survival.

Importantly, these instant reactions, with strong surges of energy, were not voluntary, nor carefully thought through and meticulously implemented. Natural selection favored those who reacted automatically and involuntarily. Thus, the evolution of the sympathetic nervous system (SNS), our built-in survival mechanism.

Today, the SNS is at the ready at the blare of a car horn, a voice raised in anger, a vicious dog heading one's way, and the starter's pistol or whistle at the start of a race. In our modern lifestyle today, it also may be activated in anticipation of stressful situations: an interview, an exam, a speech or presentation to be made, or an argument with a spouse, with an associate or a stranger. *Fight or flight* responses may be initiated by losing one's temper, for example experiencing "road rage" when the other driver cuts you off and with only your hard braking to prevent a collision.

Here's what happens when a situation invokes a SNS response:

The SNS initiates an immediate surge of hormones. One is adrenaline, which raises the heartbeat to deliver more oxygen to the muscle cells. The lungs are prompted to breathe faster, delivering even more oxygen to the blood to increase energy in the muscles and increase alertness in the brain. In addition, a surge of glucose enters the bloodstream to further boost energy and responsiveness. This action takes place in seconds, without the conscious actions or even the awareness of the individual. At the same time, the vagus nerve triggers the digestive system to slow down or come to a full stop so that the energy used in peristalsis, the contractions that advance food through the stomach and intestines, can be diverted for more urgent, immediate needs.

Now the body is at its readiness for peak performance. The individual is running on surges of oxygen, adrenaline, and glucose. Alertness, strength, speed are maxed out. Given that situations can change, ending the need for fast action and high attentiveness, and further given that these surges of high energy can quickly deplete energy stores, the ANS recognizes the need to come down and switches over to the parasympathetic nervous system.

The Parasympathetic Nervous System

While a number of ANS stimuli are engaged to initiate and orchestrate the action-oriented SNS response, a very different process takes place behind the scenes to reverse the condition and bring things down. Now, a ubiquitous, extensive component of the ANS takes over. It is the 10th cranial nerve, the ventral or front part of the vagus nerve, extending from the brainstem between the brain and the spinal cord to the key organs and muscle groups. Mediated by the ventral vagus nerve, a series of actions counteracts and deactivates the extreme reactions of the SNS, leading to what is often referred to as the rest and digest phase. Blood pressure and heart rate slow to normal levels, breathing slows, and digestion resumes. The body assumes a state of equilibrium, also called homeostasis.

The heartbeat is regulated by one branch of the vagus nerve that connects with the heart at the top right of the myocardium, or heart muscle, at the point just above the right atrium, at a nexus or cluster of nerves called the sinus node or, more formally, the sinoatrial node. Here the heart rate receives a first controlling impulse, slowing down to 100 or fewer beats per minute. The electrical impulse from the sinus node then descends to the atrioventricular node, then a nexus called the bundle of His, then to the right and left bundle branches, which extend to the right and left atria and ventricles, and finally, the Purkinje fibers, all moderating the heart rate and transmitting contraction signals to the right and left sides of the heart muscle.

Other branches of the vagus nerve extend to the lungs and diaphragm to slow breathing and resume deeper, more relaxed breaths. Another branch reaches the esophagus, trachea and on to the gastrointestinal system, allowing peristalsis contractions to resume, pushing food through the stomach and into the small intestine where digestion and assimilation continue normally. The vagus nerve continues through the entire digestive tract, including the large intestine and the colon. The energy that had been diverted for sympathetic nervous system action and alertness now returns to performs its usual metabolic functions.

Ventral and Dorsal Vagal Complexes

The Polyvagal Theory divides the vagus nerve into separate functional areas, based on responses to situations, from stressful to non-threatening. It postulates that the responses are based on environmental clues, from extreme, obvious and immediate existential threats to interactive, social engagement opportunities prompted by subtleties such as eye contact and facial expressions.

These responses are based on the recognition that the vagus nerve has two distinct branches, the more primitive dorsal branch, and the more advanced ventral branch. Between these branches, the vagus nerve influences and activates a primitive shutting down in the face of extreme threats and inescapable dangers, and it also initiates the parasympathetic response to calm and relax us, but also to go beyond passivity and facilitate communications and social interaction.

The dorsal branch of the vagus nerve evolved first and is found today in most vertebrates. It is called the vegetative vagus because it describes a primal instinct, especially in reptiles, amphibians and primitive primates. It is what we refer to as an animal playing dead when threatened and unable to escape. The animal's heart and breathing functions slow dramatically. This is not a bluff, as the creature can be truly immobilized. In the extreme, the slowed heart rate can enter the state of bradycardia—extremely slow heart rate—causing apnea, shock, and potentially death.

In humans, primitive dorsal reactions are far less common than among reptiles and other vertebrates, but nevertheless, can exist in certain extreme situations. For example, when an athlete greatly overextends capabilities and the heart suddenly slows or goes into cardiac arrest. Other, less extreme but serious human dorsal reactions can include speechlessness, freezing in place (unable to move), involuntary urination or defecation, disassociation and disorientation, fainting, and shock, which can become a potentially lethal circulatory shutdown.

The ventral branch of the vagus nerve evolved in more complex mammals and given its diversity of functions, is called the *smart vagus nerve*. It is the ventral branch that initiates the parasympathetic response to counter or reverse the fight or flight sympathetic nervous system response to danger or stress. It lowers heart rate, slows breathing, and also enables the individual to consciously practice self-calming, enabling recognition that the situation has returned to secure.

On a more fundamental, ongoing basis, the ventral branch mediates a broad range of activities, including eating, nursing, kissing, speech, singing and eye contact. In these and other contexts, the ventral branch encourages and facilitates communication and contact between people, which is considered to be a form of assuring safety. This enabling of two-way social interaction can be credited with the successful evolution of the Homo Sapiens species to the detriment of other less communicative species, including Homo Neanderthal, and perhaps the lesser known, recently discovered Homo Denisovan species.

The role of the ventral vagus nerve branch in facilitating eye contact extends in Polyvagal Theory to treatments for autism, and this will be addressed later in Chapter 9. In summary, the Polyvagal Theory hypothesizes that facial gestures, including smiles, eye openings, relaxation of the facial muscles, eye contact, and body language can initiate physiological responses in others. This has led to successful treatment of autism patients, especially children. When this hypothesis was tested, favorable results were achieved. This is believed to be the awakening of social engagement among the autistic. For the first time, facial expressions were exhibited by autistic children, including smiles and maintaining eye contact, as well as verbal expressions.

These findings encourage wider applications in stimulating social engagement among those suffering from depression, loneliness syndromes, and hard-to-treat cases of PTSD. There may also be non-medical applications, ranging from negotiations to advertising, involving software and AI applications that can analyze facial expressions of opponents in a negotiation or debate or prospective customers. Conceivably, electronic ads at or near the point of sale can adjust their messages instantly, based on the facial expressions of people seeing the ad, having determined a state of acceptance or rejection.

Other applications can include having reassuring facial expressions easily visible when needed to provide reassurance to persons having fears or anxieties. For example, to help someone overcome fear of flying, a companion can be trained to show a smiling, nodding, open-eyed expression to the person at key moments (e.g., doors closing, taking off, landing). Alternatively, apps could be developed for smartphones and tablets showing encouraging faces and issuing reassuring sounds. Flight attendants are trained to be reassuring and their warm expressions and soft-spoken words are now understood to have a physiological basis for being comforting.

Chapter 2: Communication and Connection

Communications had traditionally been understood to be the transmittal of content from point A to point B, or to multiple points. As the technology of communications evolved from primitive signals, including hand and arm motions, and limited spoken language, to more elaborate and farther reaching transmissions, communications became more complex, more informative, and better able to influence and potentially motivate large groups of people. But the concept of communications remained primarily a one-way street, going from a single source, outward, to reach one or many. With the exception of face-to-face meetings, there was no response or feedback mechanism available to message recipients as they could listen or read, but could not reply.

In more recent times, this has changed to a two-way street and even more recently to an almost infinite number of streets, metaphorically speaking, as those to whom the communication is directed can not only react, but may now respond. The responses can range from simple replies to a text or email that may reach the sender and a handful of "cc respondents," to potentially thousands (or even millions), when the messages are exchanged via social media. Feedback is taken for granted today. It's easy, it takes minimal effort, and it can be almost immediate. Even a simple Like or click on a rating, 1 to 5 stars, for example, is a form of feedback.

But there are concerns that while these technological advances in communications are making broad-reaching feedback possible, they are denying us the advantages of face-to-face interpersonal contact. New theories, notably the Polyvagal Theory, are identifying subtle facial and bodily cues that can provide us with knowledge of true intentions, whether the person we are facing is a friend or foe, whether we can feel secure and safe, or if we should acknowledge a potential threat. Some interpret this perspective as a call to return to the personal and to reduce the dependency modern society has created on technology. Polyvagal Theory is testing and confirming the relationship-building and communications that are uniquely available in face-t0-face environments.

Social Communication Systems

Social communications, at its most fundamental level, is an interaction between two or more individuals that invokes a response or reaction at some level. The response can be a barely conscious or subconscious acknowledgement of a smile or any other facial expression, or a gesture, or mannerism. Moving up the scale of complexity, social communication can include conversation, mutual engaged eye contact, and on up to degrees of information exchanges and learning. It may be contended that the highest levels of social communication can include relationship-building, attitudinal changes, and persuasion to influence behavioral changes.

The Polyvagal Theory has introduced concepts of bodily reactions being initiated by social communications, including simple facial expressions.

The fundamental model of social communications includes these interactive steps, which involve a message, this message can be of any level of simplicity or complexity:

- A sender conceptualizes a message to send to someone

- The sender encodes the message for transmission
- The message is sent to a recipient
- The recipient receives and encodes the message
- The recipient acknowledges the message

Messages can be any form of information, encoded and transmitted as a smile, a facial expression, making eye contact, a greeting, or any written form, from texting to email. Encoding or interpretation can be perception of the message, e.g., seeing the smile, reading the text, and acknowledgement may be returning the smile, maintaining eye contact, or responding to the text. From these initial social communications interactions, relationships can evolve, and according to Polyvagal Theory, physical or visceral bodily responses can occur.

Social communication has also been described as either interactive or transactional:

> **Interactive Approach:** This is a simple model in which a speaker directs a message to a listener, the listener then becomes the speaker, and responds to the originator who now plays the role of listener. This is compared to a game of ping-pong or tennis, with the single ball going back-and-forth in logical sequence. It describes a polite

conversation or exchange of text messages, for example.

Transactional Approach: This is a more complex model in which the messages do not go back-and-forth sequentially, but simultaneously, with multiple transactions involving speaking and listening. This describes real world communications, as it is no longer a simple, polite conversation, but a multitude of messages talking over one another. For this to be successful, the communications would need to be recorded in some manner for subsequent review and interpretation, as the messages have been richer, more informative and potentially more engaging, but in being so, are potentially overwhelming.

Dr. Steven Porges, creator of the Polyvagal Theory, has expressed concern that technology is interfering with the social communications he considers essential for the interactions that build social relationships. In effect, even though texting, for example, is almost instantaneous, it is highly simplified, without the vast complexities of signs and signals that face-to-face communications can entail. Porges says "our systems crave the reciprocity and synchronicity of face-to-face interactions." (Porges, S. 2016). According to Porges' Social Engagement System, healthy interpersonal engagement depends on cues, such as facial expressions, eye contact and vocal inferences. These give us a sense of safety as the other person is presenting positive, encouraging, inviting signs, and not exhibiting cues that might suggest risks and danger, such as previewing aggression and hostility.

Porges calls these unconscious responses to things in the environment, from reaction to physical threats to subtle signals from people, as neuroreception and while the sympathetic call to action has been understood for some time, it is from close study of the calming sympathetic responses that Porges has linked the emotional to the physical.

Similarly, we can transmit these signals when trying to put others at ease. The key point is that the cues are not necessarily voluntary, but occur automatically as the vagus nerve interprets our intents and emotions, and translates them to physical manifestations and the facial and vocal cues that put others at ease. This is the parasympathetic nervous system response in action, calming, relaxing, slowing key bodily functions.

There is concern, as expressed by Porges and others who study the intricacies of our psychological and physiological responses to subtle, sometimes intangible cues that we send and receive. The concerns are that technological advances in the means of communications are diminishing face-to-face, human interpersonal exposures that enable the bodily responses that are essential for our safety, comfort and trust of others.

Social Engagement Systems

The Polyvagal Theory describes a new defensive circuit that is the newest to evolve, a mammalian vagal pathway that is going from the brainstem to the heart, and also to the nerves that regulate the muscles in the face and head. The Polyvagal Theory in the social engagement system is the neural-regulation of the muscles of vocalization, the muscles of listening, the muscles of facial receptivity, and the muscles of gesture. Since these are linked to the nerve regulating the heart, it becomes a co-regulating system. (Porges, S. 2016, Podcast)

Within the context of the Polyvagal Theory, a social engagement system is a two-way exchange between two persons, involving both expressions and receptions of the messages implied by those expressions. These interactions create what is called mammalian attachment, and occur primarily in the facial area: the eyes and ears, the larynx and mouth, and extending, to a lesser extent, to the body down to the region of the diaphragm. All 12 of the cranial nerves extending from the brainstem participate in these expressions and reactions, with the 10th, the vagus nerve, having the most influence. Combined, this is what is commonly referred to as body language.

According to the Polyvagal Theory, these expressions and reactions initiate physical changes that facilitate social interaction. For example, in response to an open expression, a smile or a softly or moderately spoken voice, the respondent's ears may actually open wider to receive the words, and the middle ear muscles contract, reducing or blocking outside noises. The smile being perceived may be reciprocated, eyes may correspondingly open. Safety, security, absence of danger are subconsciously perceived.

Another, less immediate response is digestive, which is why taking a meal together can have a bonding effect. It is more than the pleasant atmosphere since the digestive system is being relaxed by the social interaction, by the facial expressions and vocal tone. The two people at lunch are putting each other at ease, albeit unconsciously. The opposite effect, of course, would be a confrontational meal, during which the digestive system might shut down in consequence to a stressful reaction. Then, there would be a negative response, weakening any potential bond between the two.

The social engagement system can be applied under a wide range of situations. A person who has fear of flying may not be calmed by the usual presentation of safety statistics and seeing other passengers being calm. But if a person who is trusted is able to smile, show an encouraging and reassuring expression, the social engagement system may be activated, and while the nervous person may not be aware, tension and stress levels are being reduced. Flight attendants who are trained to calmly connect well with people are often successful just by speaking warmly and softly, smiling, reassuring with open expressions. Tests suggest that optimum application of the social engagement system to calm a nervous airline passenger would involve presenting a reassuring face to the passenger at key moments, such as door closing and takeoff.

Social engagement may be better appreciated when seen in its wider context, beginning with the concept of community.

We live in an age of community, but with new definitions and applications. What was once used to describe a village or neighborhood, the term community has now expanded to social engagement systems. A community can now be comprised of members of a church, synagogue or mosque, members of a local sports club, or fans of a national sports team. Radio stations may try to build a virtual community to increase the loyalty and interactive participation of its listeners, political action groups can be perceived as communities, and attract members who are committed to the community, performing interaction by making financial contributions, devoting time to the group's activities, or advancing the group's mission through solicitation of contributions or by contacting members of Congress.

Communities may also form within residential environments. For example, a private residential development is, in fact, a type of community as its resident members have certain interests in common, like security, maintenance, and provision of amenities. There may be opportunities or obligations to participate in some way in the community, such as paying common charges, or attending barbecues and other social activities. Even a single apartment building or cluster of townhomes can form their own communities, in the interest of stimulating social interaction among residents. In the extreme, a private residential development may be constructed as a gated community, open only to residents and their guests.

What all of these types of communities have in common are the qualities of social engagement groups.

Most definitions of social engagement require membership or at least a sense of membership in a group (social), and further imply an active or participatory role within the group (engagement). Other definitions include the building of relationships within the group and various interactions among group members. Social engagement is distinguished from social networks by virtue of the interactivity. In a social network, one may be a member of a group, but interactivity and relationship-building are not prerequisites. Another distinction is from social capital, which refers to the resources that the group may possess. Social engagement groups may be any size, from several neighbors or a half-dozen members of a local garden club, to multi-million members of a political action group or a proactive environmental movement. As long as there is relationship-building and interaction, it qualifies as social engagement. Participation in the activities of the group can increase its social capital and further build its social norms, which define such parameters as the social group's mission, objectives, codes, rules of membership and schedules of activities.

A primary example of a large-scale social engagement system is religious affiliation, especially when it includes regular attendance and participation in weekly services. The congregation of a church that meets on Sunday mornings, that sings, prays and discusses matters together qualifies as a social engagement system. The interaction between members of the congregation can be constructive, as in planning church activities, voting for positions of responsibility, welcoming new members, forming and training a choir, and conducting regular discussions of religious and secular matters. But the interaction may be divisive, as in cases when religious dogma creates arguments of spiritual beliefs or when the congregants disagree as to the role of the church in the larger community, e.g., the town or village where it is located. There are often disagreements within religious social engagement systems concerning involvement, for example, in political or educational affairs.

Social engagement networks form naturally within educational environments, where faculty and parents may form a Parents Teachers Association (PTA). The common connection parents and teachers have with the school creates the basis for a social group, as the meetings, exchanges of ideas, forums and hearings are all forms of social interaction. It is important to note that not all members of the social engagement group have to build relationships or participate actively. As long as enough members are sufficiently active to make decisions for the group, stimulate change, disseminate information, and keep the group vibrant and current, it qualities as a social engagement group or system.

An important aspect of social engagement is improved health, both mentally and physically. Many studies in the U.S., Europe, and Asia have concluded that social engagement and interaction are beneficial to person's well-being. Among the younger and middle-aged, social engagement is found to encourage more positive, optimistic outlooks, stimulate learning and experimentation with alternative lifestyles. Among seniors, social engagement has been found to slow the onset of dementia and other cognitive disorders associated with aging. Reasons for this effect trace to greater stimulation from group activities, including remaining more physically active and being more mentally challenged. But referring to Polyvagal Theory, the real benefits may accrue to person-to-person interactions, stimulating calm and a sense of safety. This may help explain why married couples tend to live longer than individuals who live alone.

Positive effects of social engagement for older people are reflected in advertising today for senior living communities that offer group activities ranging from walking and hiking to cooking classes. Commercials show groups of people gathered at barbecues, sushi suppers, and swimming activities. All members of the community are interacting.

There are negative possibilities to participation in social engagement groups. For example, if the objectives of the group are not positive in nature. Gangs, despite all their dangerous, criminal aspects, qualify as social engagement groups, having passionately devoted members, high levels of interaction between members, and action-orientation. This may explain the challenges faced by law enforcement and public officials in dissuading gang members from remaining within their gangs. They may identify with their fellow gang members—their community—more than any other group, including their own families.

A less worrisome but potentially risky aspect of socially engaging activities can be overcommitment, wherein some members, generally a small number, end up taking on more social group obligations than they can handle. In consequence, instead of experiencing the enrichment of human contact, building personal relationships and feeling a sense of belonging, these overextended individuals experience stress, frustration and may disengage from the group.

Polyvagal Interactions

The term Polyvagal is derived from the Greek word polus or many, and vagal, pertaining to the vagus nerve. The theory was first presented by Dr. Steven Porges, then at the University of Illinois, when he was directing the Brain-Body Center for research. The Polyvagal Theory identifies sensitivities and interconnections between the brain and visceral organs, notably the head and face, the heart, lungs and digestive system. The autonomic nervous system is networked or connected with these body parts and is susceptible to their impulses as they travel to the brain. These body-toward-brain signals are called "afferent influences." According to the theory, the vagus nerve initiates the action-generating parasympathetic nervous system, stimulating the heart and lungs, and suppressing the digestive system to conserve energy for more immediate needs. It also can activate the face, eyes and ears, the mouth and larynx, where facial expressions are created and managed.

As investigations into the Polyvagal Theory continued, new hypotheses and ideas for its application emerged. One of the most promising aspects is the correlation between facial expressions and body, or visceral, reactions. Extending this to interpersonal relationships, work is underway to engage Autistic patients who otherwise have not been responsive to current, accepted treatments. There is evidence these Autistic patients can relate physically and emotionally to facial expressions of others.

Returning to the discussion of communications, the overwhelming presence and influence of technically-enabled communications, even with its easy, immediate response and feedback mechanisms, is depriving individuals of the face-to-face contacts that provide the rich opportunities for influence, security, calm and all other aspects of human interaction. Now, with the guidance of Polyvagal Theory, these interactions may transcend simple comforting gestures, and actually influence bodily reactions. These responses are mediated by the vagus nerve, including visceral activities, like heartbeat, respiration, facial gestures and reactions, and digestion.

When extreme, threatening situations occurs, social responses can be mediated within the autonomic nervous system, following three separate types of response. The most extreme response to stress and threats results in a shutting down the person to the extent that any action is not possible. The person freezes in place, unable to move, or think, or speak. Terms like dumbstruck and speechless describe this extreme reaction. Fainting, involuntary urination, and going into shock may occur.

A more normal, but still dramatic response is the fight or flight reaction, reducing or mobilizing the visceral response, which initiates suppression of metabolism in conditions of stress or danger, and controls the activation or slowing down of the digestive functions. Energy is conserved for action, and hormones, including adrenaline are released. This is the well-described sympathetic response.

A third phase engages the parasympathetic nervous system, which the vagus nerve activates to calm or slow down the fight or flight sympathetic nervous system response, as well as resuscitating the person who experienced the extreme shutdown response and froze, fainted or went into shock.

These responses have been studied for some time. The role of the central nervous system in regulating visceral organs, like the heart, was recognized during the 19th century. As early as 1872, Charles Darwin maintained that there was a direct relationship between the heart and the brain, as the heart is somehow agitated or otherwise affected, there is a reaction in the brain, and conversely, any reaction in the heart, such as stimulation, is transmitted to the brain by what was then called the pneumogastric nerve, which we now call the vagus nerve. These early discoveries laid the foundation for research correlating the brain and bodily organs, research which continues to advance new discoveries today. The Polyvagal Theory was postulated first by Dr. Porges in 1994, and its implications continue to be explored and validated.

Chapter 3: The Physiological Regulation of Emotion

When we experience something upsetting, we react, emotionally and physically. The emotional reaction may be anger, or anxiety, or stress. Or it may be fear. As has been discussed regarding the sympathetic nervous system's fight or flight response, in more stressful or fearful situations, our body may quickly prepare for action. In less threatening situations, our emotional reactions may trigger more moderate responses, such as a slightly elevated heart rate, slightly faster breathing, no real impact on our digestive system. In the facial zone, our pupils may dilate slightly, our hearing may become slightly more acute, our facial expressions may show only a small degree of frustration or disappointment if the situation so justifies.

But whether the reactions are extreme or moderate, they may be brought under control, bringing us back to a state of normalcy, or homeostasis. According to Polyvagal Theory, the control may be mediated physiologically. Bodily actions affecting emotions, which, in turn, affect the physiological reactions or overreactions to stress. In other words, emotional responses first trigger physical reactions, then physical activities calm the emotional reactions, bringing everything back to normal.

This chapter will provide practical examples that can be applied to a range of emotionally-driven physical reactions.

Managing Stress

Stress. Everyone knows—or thinks they know—what stress is, and the result is a variety of definitions. Some mix stress with anxiety, others think stress is tension due to anticipation of a difficult challenges. Still others perceive stress on a more physiological basis, including an elevated cardiovascular and respiratory state. Whatever the cause, and whatever the definition, there seems to be general agreement that stress is an undesirable and potentially destruction event that should be avoided, or neutralized when it does occur. The goal of stress-avoidance or stress-elimination is to achieve a state of calm relaxation, that being the antithesis of stress.

Prior to the emergence of the Polyvagal Theory, the autonomous nervous system was understood to embrace two physiological stress response mechanisms:

> The sympathetic response, which is the well-known call to immediate action when danger or high risk creates a high level of stress. This leads either to fear, panic, and preparation for escape or evasion or it leads to anger, rage, and preparation for aggression.

> The parasympathetic response deactivates the sympathetic responses elevated levels to calm and cool things down, lowering elevated heart rate and slowing rapid breathing, returning the digestive system to normal and the person to an overall peaceful, relaxed state of mind. This response is mediated by the ventral vagus nerve, meaning it is located towards the front of the body.

But when Dr. Porges introduced the Polyvagal theory, he added a third level of reaction, the most extreme, when the parasympathetic nervous system causes a near-total shutdown of key bodily functions:

> This third level is called the dorsal vagal, meaning it is found in the rear part of the body. This is the

response that may cause an animal to play dead or a person to become immobilized, incapable of movement or rational thought. A person may become speechless, enter deep depression, faint or go into shock. In the most extreme situations, cardiac arrest may occur, leading to death if not reversed immediately.

In almost all stressful situations we may encounter in our daily lives, the levels of stress are perceptible, perhaps somewhat debilitating, with our sympathetic responses causing moderately elevated heart and breathing rates, some sweating, and a feeling of nervousness or anxiety. If the event or situation causing the stress can be completed in the near term—the interview is over, the test completed, the dangerous road safely crossed, the steep hill easily skied down—the calming parasympathetic response will automatically bring you down, cool you off, put you in a state of relaxation, possibly elation if the challenge overcome was formidable.

But if the stress cannot be neutralized quickly by resolving its cause, can it be brought down consciously and deliberately? A state of relaxation can be achieved through certain practices that have become better understood through the Polyvagal Theory.

The objective of stress reduction is called the *relaxation response.* It is the body's physical actions, mediated by the ventral vagus nerve, that stop the surges of adrenaline, slow heart rate, lower respiration, and induce a state of calm.

Most techniques to induce the relaxation response center on the control of breathing. When under stress, our breathing becomes more rapid and more shallow. This is an involuntary sympathetic response, but it can be reversed, bringing down the sympathetic response and swapping in the calming parasympathetic response.

Polyvagal Theory originator, Dr. Steven Porges, recommends tricking the body by assuming deliberate control of the breathing process. Find a quiet place where there are minimal distractions. You may sit or stand, and in either case, be comfortable:

> Begin by telling yourself it is time to become at peace with yourself.
>
> Consciously become aware of your breathing.
>
> Begin to take slow, deep breaths, pausing for a moment or two between each inhale and each

exhale.

Your breaths should be deliberately deeper than usual, as you inhale, extend the diaphragm outward, and exhale fully, pulling in the diaphragm. Remember to pause briefly at each cycle.

Count the inhales, starting from one. alternatively, count backward from 10. Repeat after each ten-count. Options: count exhales, or count both inhales and exhales.

Keep your mind on your breathing and nothing else. As other thoughts try to intrude, do not become angry but gently refocus your thoughts on breathing in deeply, pausing, breathing out deeply.

Remain conscious of your efforts to force the diaphragm outward on each inhale, and inward, toward your spine, on each exhale. Be aware of the pauses and be aware of the count of each breath.

Continue the deep, deliberate breathing for a minute or two, ideally for five to ten minutes if time and patience permit.

When you resume normal breathing, keep a positive outlook and do not allow yourself to return to stress-inducing thoughts and worries. The best way to keep the stressful thoughts away is to tell yourself you are calm, in control, at peace. Smile, as smiling is a ventral vagus nerve response and will further induce a warmer, more relaxed physical and mental state.

Traditional meditation is a more formal method of achieving the relaxation response. Functional Magnetic Resonance Imaging (fMRI) studies of brain activities and responses confirm that mediation practiced over time can induce physiological changes in brain function, long term meditation practitioners believe they have gained control of their emotions, and have greatly reduced the onset of stress and its bodily reactions. The fMRI tests show increases in blood flow to parts of the brain being activated by the meditation, confirming the onset and continuation of the relaxation response, mediated by the ventral vagus nerve system.

Meditation may be performed in a manner similar to the above breathing exercises, but with the person seated comfortably in a quiet environment, eyes closed, and full focus on either the breathing, or a mantra, which is a humming sound issued on each exhale. To some people, the mantra helps clear away extraneous, distracting thoughts.

Mindful meditation is a Buddhist-inspired variation in which the individual, while in the relaxed pose, remains conscious of the moment, aware of every feeling, every sound, every other sensation. It is continuous, dispassionate, and non-evaluative state of mind. Mindful meditation followers believe their awareness of their total incoming environment, attention to the present moment, keeps their minds clear of other thoughts, leading to deeper relaxation and elimination of stress and anxiety.

There are apps available online, free or at low cost, that can help the meditative process, providing soft, peace-inducing words and suggestions, and either restful music or sound effects, like waves crashing, gentle rain, crickets chirping, and wind blowing.

What about Yoga? The poses and stretches of Yoga are effective self-regulatory approaches that are covered in the following chapter, which covers physical vagal applications. While Yoga incorporates a range of physical activities that stimulating toning of the vagus nerve, the deep, deliberate breathing exercises just discussed are key components of the Yoga routines.

Overcoming Depression

Depression is a state of mind. It is mental, yet it is also a physically-affected disorder. The Polyvagal Theory establishes that depression is caused by malfunctions of the vegetative nervous system, the action-directing sympathetic nervous system is keeping the depressed individual in a continuing state of low level stress. The calming parasympathetic nervous system may be functioning, but not sufficiently to fully counteract the sympathetic responses. The depressed individual falls into a state often characterized by apathy, lack of drive and motivation, and a general lack of energy. Yet, despite the lack of energy and drive, the condition inhibits the ability to rest, and sleep become lengthy but not restful. The agitating results of stress remain behind the scenes, keeping the action impulses from fully slowing down and dissipating since the parasympathetic response is only partially successful in its calming effects, and heart rate and breathing rate are only partially slowed. The result is a limbo between action and inaction, expending energy while not allowing relaxation to be achieved. Depression can worsen as the situation appears endless, hopeless, with further weakening of resolve as energy reserves continue to be dissipated. There is little interest in social activity and empathy toward other people.

Breathing, as we have just seen with management of stress, may play an important role in reducing or eliminating depression. The reason breathing can be so influential is due to the vagus nerve, which plays a key role in our breathing rates and depths. When the sympathetic nervous system causes the heart rate to increase, and breathing rates to increase while making breaths more shallow, these vagus nerve-instigated reactions to stress can be managed. Instead of allowing the vagus nerve to affect breathing, we can use breathing to affect the vagus nerve, inducing it to switch modes from sympathetic to parasympathetic. This is referred to as *putting on the vagal brake*. The two phases of breathing, inhaling and exhaling, are involved.

As with stress reduction, the technique to manage depression involves taking deeper, slower, more deliberate breaths. A deep inhale signals the parasympathetic vagal system to engage and stimulate relaxation, and it also increases the amount of oxygen that is inhaled and delivered via the hemoglobin in blood, to the cells, where its metabolism results in greater energy output. There is also a purely psychological effect, as the deliberate, forceful deep breath gives feelings of strength and resolve, building self-esteem and trust.

Forceful, slow exhales continue the relaxation response, exerting an influence on the vagus nerve to further slow things down, getting the sympathetic response to back off. Psychologists believe that the slow, forceful exhale adds a sense of accepting life and feeling better overall.

There can be immediate results, within minutes of practicing deep, slow, deliberate inhales and exhales. This procedure may be further enhanced with the meditative and mindfulness procedures cited for stress management. Depression may also respond to Yoga and other activities that combine physical efforts in harmony with the deep, managed breathing exercises.

Importantly, depression is a serious disorder that may require extensive psychotherapy and medication, and should only be diagnosed and treated by qualified medical personnel. Further, the effectiveness of the breathing exercises described here depends in large part on long term continued practice, as one brief session of deep breathing may bring momentary relief of depression, but daily repetition, over time, is necessary for longer term diminishment of symptoms.

Depression, epilepsy, Alzheimer's disease, cluster headaches and irritable bowel disease are also being treated by electrical stimulation of the left vagus nerve. A device is implanted subcutaneously, connected to the vagus nerve by a thin wire. Signals are sent through the vagus nerve to its point of origin in the brainstem, and on to the parts of the brain affecting depression or other diseases. Impulses may be triggered when disturbances are detected. Results of electrical stimulation for depression, epilepsy and other diseases have been partially successful so far. Recent introduction of non-invasive electrical vagal stimulators encourages expectations for wider applications.

Asperger's Spectrum and Autism

Application of the Polyvagal Theory is based on the professional observation that patients, especially children, who are affected by sensory-processing disorders that fall under the Asperger's Spectrum, are frequently characterized by a lack of eye contact with others, and general failure to communicate. Polyvagal Theory connects facial expressions, including eye contact, with actions mediated by the vagus nerve.

Polyvagal Theory states that the social, or myelinated part of the vagus nerve connects with the heart, lungs and digestive system—this is well-known—but also to the facial muscles, the eyes and the ears. These facial region functions comprise the elements of facial expressions, which turn out to be key in reaching and motivating Asperger's Spectrum patients.

It was observed that these patients, especially children, could not process sounds, resulting in covering their ears or turning away when spoken to. Another observation was the expressionless facial poses that are typical of Asperger's children. It was determined that stimulation of the children's social vagus nerve could help them received sounds more normally and be able to express themselves facially. In addition, many Asperger's Spectrum patients have difficulty controlling their breathing.

Work to apply the Polyvagal Theory to improving Asperger's patients has been performed by chiropractors, who, contrary to many misconceptions, are nerve system doctors whose specialty is helping nerve systems to rewire themselves. They are at the forefront of helping patients, from those with mild Asperger's symptoms all the way through the spectrum to autism. Chiropractors follow a salutogenic model of healthcare, the goal is to optimize performance and increase the state of healthiness.

Chiropractic doctors working with Asperger's patients recognized that Polyvagal Theory is compatible with their salutogenic healthcare model. Specifically, it supports the concept that the body can self-regulate itself, and it can self-heal under the right conditions. This inspired recognition that the Polyvagal can enable the doctors to tap into the healing potential. They apply neurological exercises that stimulate vagal tone in their patients, and this gives the Asperger's Spectrum patients new ways to hear, perceive, respond to people and situations, smile, speak, and maintain eye contact.

One form of neurological exercises is atlas adjustments, or Vertebral Subluxation, the manipulation of the spine to correct vertebrae that have become misaligned. This procedure benefits from the close proximity of this part of the spine with the vagus nerve.

The other form is manipulation of the vagus nerve in the neck, accessible as it exits the jugular foramen, which is an aperture, or opening, at the base of the skull.

Patients on the spectrum report achieving a greater awareness of their bodies as a result of Polyvagal Theory-inspired manipulative exercises. This work is early in its development but shows promise for continued progress.

According to Polyvagal Theory author, Dr. Porges, there is a direct, positive link between the autonomic nervous system, which controls our organs' responses to danger and stress, and challenges in learning and socialization. This is most evident with Autism patients. Porges has postulated that we are social creatures and look to each other for comfort, security, a sense of belonging. But autistic people are unable to perceive or recognize these safety cues, and are therefore unable to relate easily to others.

Porges and others have developed an auditory system that is able to reach the autistic and ease their discomfort with social and interpersonal relationships. Computer-generated vocal sounds and intonations were created, and the sound is described as similar to a mother's lullaby. Studies beginning in 2014 confirmed that when listened to for one hour over five consecutive days, there were measurable improvements in auditory processing and increased vagal control of the heart rate. The result was more open, receptive, communicative autism patients (Porges, S., 2019).

Chapter 4: Self-Regulation Vagal Applications

The Polyvagal Theory identifies three important physiological events:

> First is activation of the sympathetic nervous system, which can dramatically raise our energy levels, elevate heart and respiratory rates, and put us in a state of readiness for action. This is the fight or flight response, evolved early in our evolution as a protective mechanism.
>
> Second, activation of the parasympathetic nervous system, the ventral vagal, social engagement response, which acts to deactivate the state of high energy and readiness, returning heart and breathing rates to normal, enabling us to achieve calm and relaxation, known as the *rest and digest* stage.
>
> Third, activation of the dorsal vagal freeze response, which is an overwhelming reaction (or overreaction) of fear, panic and a sense of no

escape, which may extend to a total shutdown, immobilization and shock.

These are all functions of the autonomic nervous system and happen involuntarily and automatically in response to external stimuli. But Polyvagal Theory provides a roadmap of voluntary action we may take to help the parasympathetic response come into play when we want. By engaging the vagus nerve with certain actions, it is possible to induce calming, relaxation, a sense of peacefulness and hope.

In modern life, the majority of stresses we experience are not life-threatening, as brief frustrations, insults, fears, tensions, disagreements that, while minor, nevertheless induce a sympathetic response that can put us on edge, raise our pulse rates and blood pressure levels and put us in a state of anxiety. It is to counter these responses and bring us back down that the following self-regulation vagal responses are proposed.

Yoga Poses and Stretches as Therapy

The practice of Yoga has been traced back to early Buddhist teachings, and today enjoys a worldwide following. It is now known, thanks to Polyvagal Theory, that certain Yoga movements, stretches and poses or positions, also called asanas, actually stimulate the vagus nerve, creating vagal tone and inducing the calming, relaxing parasympathetic response. There are parallels between Yoga and meditation, with common focus on conscious, managed deep breathing, and working to concentrate on the present moment by avoiding the intrusion of outside thoughts.

Yoga is often practiced in group sessions with an instructor leading and demonstrating each movement and pose. Having an instructor can have the added value of coaching, correcting movements and poses, and also voicing reassuring phrases. If you are a beginner, or have had only limited experience with Yoga, be sure to select a beginner's-level class. The instructor will be able to work you slowly through each exercise, as you do not want to start in an intermediate or advanced class and find yourself falling behind or not knowing what everyone else seems to be doing effortlessly.

While an instructive class is recommended for beginners, it is possible to begin Yoga training individually, at home or at any place of calm, quiet and privacy. No equipment is required except for a Yoga mat, typically a rubberized or foam cushion to protect the back, knees, and joints.

A mat may not be necessary on a carpeted surface.

Yoga instructional videos are readily available online and can provide demonstrations that are better than written instructions. However, to get started, the following are five of the simplest, most popular Yoga poses you may try. In every pose, be sure to concentrate on the movement and stretching, and be conscious of your breathing. Engage and tone the vagus nerve by forcing your diaphragm outward on the inhale, then contracting it by pulling in toward your spine on the exhale.

> **Mindful Breathing:** This is a simple starter that initiates the essential deep, diaphragmatic breathing pattern. Lie on your back, legs extended, and arms crossed on top of your lower abdomen. (Alternatively, one hand on the lower abdomen, the other on your chest.) With eyes closed, or while staring at a point on the ceiling, begin breathing slowly and deeply. Inhale fully, pause for a few seconds, then slowly and fully exhale. Take your time, the exhale should take about eight seconds. Feel the diaphragm rising and lowering with each breath. Keep this up for a minute or two to set the mood for the next poses. As you proceed through this pose, try to more

forcefully extend and contract the diaphragm, but do not strain.

Balasana: Also called the Child's Pose. From a standing position, bend your knees to the ground and then assume a facedown position, resting on bent knees and sitting back on your calves, forehead on the mat and arms fully extended so your shoulders are next to your ears. Try to settle fully down on your shins, and reach forward and press downward with your hands as far as you can. Feel the stretching through your back, arms, and shoulders. Be sure to breathe slowly and deliberately, being aware of each inhale and exhale. Hold this pose for 30 to 60 seconds, then slowly move to the next pose, Cat/Cow, which flows easily from the Balasana pose.

Cat/Cow: A good way to loosen tight back muscles and stimulate circulation. Rise up from the Balasana position to assume a position on hands and knees, toes pointed backward, and with your back level and parallel to the floor. Your knees should be directly below your hips and hands directly below your shoulders. Raise or arch your back toward the ceiling (like an angry cat) and hold it fully extended for a few seconds, then lower your back to its lowest position, pushing your gut as far downward as you can (like a swaybacked cow). Remember to breathe deeply and be aware of pushing out and pulling in your diaphragm. Repeat the rising and lowering cycle, fully extending each time for one minute, or longer if preferred.

Downward Facing Dog: A Yoga standard, and easy to assume from the Cat/Cow pose. Begin while still on hands and knees, and slowly, fully extend your arms, and raise your hips and rear toward the ceiling. Allow your heels to rise, and initially, keep your knees slightly bent. After 10-15 seconds, try to straighten your legs, further pushing your hips upward. If you can, lower heels so your feet are flat on the mat. Then, if you are able, walk backward on your hands, a few inches at a time, to extend the stretch of your back, hamstrings and calves. Hold for 20 to 30 seconds, then relax.

Cobra: A good pose to improve posture as well as to stimulate relaxation. Lying prone, face down, flat on the mat, legs and feet extended backward, and with your hands next to your shoulders. Inhale slowly and deeply as you slowly push your head and shoulders upward. Slowly raise the head, shoulders and upper back until you just begin to feel the strain in your lower back. Pause at this point (do not overextend or strain your back), complete the inhale, then slowly lower your upper back, head and shoulders to the mat, exhaling as you lower. Repeat the cycle eight or more times, or for one to two minutes. As an alternative, you may perform fewer cycles by holding the upward part of the post for longer periods before lowering back to the mat.

Meditative Vagal Stimulation

Meditation has been, and remains, a stalwart factor in the general field of stress management, help people achieve calm, inner-harmony, self-control, as well as building self-confidence, and overall, enjoying life to the fullest extent. Meditation began is India and Asia, and while it has been taught and practiced for centuries, today, meditation is recognized to have measurable physical as well as psychological benefits. There are many approaches to achieving a meditative state, and this section will present the best known, and will provide instructions for the reader to easily follow.

If there is a single, overriding objective of meditation, it is the elimination of random thoughts and the creation of periods of thoughtlessness. Clearing one's head, so to speak, of extraneous, intruding thoughts so that the mind is relatively clear for at least a few minutes a day. Why? Because it evokes a sense of calm relaxation that continues for hours after meditating. There is scientific proof that the relaxation and calm have physical manifestations: functional MRI (fMRI) examinations confirm that blood flow changes in the brain during meditation, and over time and with continued meditative practice. Permanent physiological changes are manifested.

Further, if there is one common element in most forms of meditation, it is conscious, managed breathing. Being aware of each inhale and each exhale, deliberately breathing more slowly, fully and deeply, forcefully extending and contracting the diaphragm.

The Polyvagal Theory supports these practices as they engage and tone the vagus nerve, inducing a strong parasympathetic response.

Here are six of the most popular forms of meditation, but no single one is better than the others. It is recommended to try several approaches and decide which is the most comfortable and effective. In all cases, give your meditational exploration adequate time as skill in attaining a true meditative state takes practice and repetition. Try to find a calm, quiet place where you will not be disturbed by interruptions, intruding noise, or other distractions. You may find that calming music or sound effects, like wind, falling rain, bird or cricket chirps, or waves crashing can help the mood. These sound effects are available online as apps.

As you will see, most of these popular techniques can be practiced by you without help. There is no need for a "guru" or instructor. However, some people are more comfortable having a coach or guide, and each individual may decide what's best.

The first two types have the common characteristic of a mantra:

> **Mantra meditation** presumes that pure silence is not enough to achieve a complete meditative experience, and advises participants to hum a sound that is calming and repetitive, like the famous ohm sound. The sound should be low but

audible, and is expressed continuously, slowly, such as during each exhale. Any sound that you feel is pleasant and relaxing should work. Mantra meditation is appreciated by persons not liking silence and finding comfort in a repetitive sound and its associated vibration felt in the vocal cords.

A potential positive to mantra meditation is that the vocal cords, which are connected to the vagus nerve, are engaged during utterance of the mantra humming sound. In principle, this can enhance the relaxation response of meditative concentration. You may feel the vibration in the throat as the vocal cords are vibrated by the low, base-level mantra chant. This may add a physical enhancement to vagal toning, which in turn may increase its effectiveness in signaling the parasympathetic calming response.

Transcendental meditation is perhaps the best known, having been introduced to the world by the Beatles, who traveled to India to take instruction from a maharishi guru, a spiritual teacher. It has become the most popular form of meditation, with a reported five million followers worldwide. It is similar to mantra meditation but is more structured in having an instructor assign a specific mantra to each person, based on the instructor's judgment and includes consideration of age and gender.

Spiritual meditation is the opposite of the mantra formats, as it is totally silent, with a near-religious sense of faith in the unknown. It is practiced with a concentration on the surrounding silence, with a sense of openness to inspiration. The practitioner seeks answers to prayers, or insights into the universe. For religious people, spiritual meditation can be a form of prayer.

Focused meditation invites concentration on the senses: sight (staring at an object), sounds, touch, and especially breathing, by being conscious of each inhale and exhale. This is considered to be good for beginners, since the concentration on breathing, for example, tends to keep the mind clear of outside, distracting thoughts.

Movement meditation is for those who prefer activity to sitting still, or who find repetitive mantras or conscious breathing to be boring or annoying. Meditation is achieved while walking, gardening, performing Yoga, or any activity that involves slow movement while allowing the mind to focus on what you are doing and avoid thinking about other things. By not requiring staying quietly in one immobile position, movement meditation may be practiced several times per day without taking off time exclusively to meditate.

Mindfulness meditation has its roots in Buddhist teachings and is more of lifestyle practice and state of mind than a specific meditation technique. In practice, thoughts are exclusively on the immediate moment and its sounds, sights and feelings, and it is easily followed by anyone who does not have an instructor and simply wants to live in the moment. Some who practice mindfulness meditation surrender themselves to the inevitable.

Auricular Acupuncture

Acupuncture is an ancient Chinese medical practice that came into prominence in the Western world during President Nixon's historic trip to China in 1971 to meet with the Communist leadership. Accompanying the president was the senior journalist, James Reston, who was given acupuncture treatment by Chinese doctors to relieve pain following emergency appendectomy surgery. Reston followed his successful experience with favorable coverage of acupuncture in the New York Times, initiating exploration and adoption of acupuncture techniques in the U.S., and a number of other countries. In subsequent years, the popularity of acupuncture in North America and Europe has risen and fallen, as conflicting reports of relative effectiveness circulated.

Today, acupuncture is recognized as a potentially effective form of alternative medicine. Various types of needles are inserted by a trained professional, who is a qualified, licensed acupuncturist, into predetermined points on the ear and body, as many of these points have bio-electric conductivity potential. According to Chinese medical concepts, health is determined and effective by qi, a flow of energy throughout the body. The qi energy impulses travel along invisible pathways called meridians.

Auricular acupuncture is limited to the ear and is prescribed for pain relief, relaxation of the mind, infectious diseases and allergies. It is also considered to be of value in treating disorders of the endocrine system and other functional and chronic disorders. It is also used to reduce withdrawal symptoms. Auricular acupuncture is distinguished from body acupuncture by virtue of its ability to be applied almost immediately, compared to body acupuncture requiring 20 t0 40 minutes of lying down before treatment. Since ear acupuncture does not require patients to undress, it can be administered to groups, which may be calming to nervous patients.

How can ear acupuncture involve the Polyvagal Theory?

Polyvagal applications include controlling irregular heartbeat, and because the vagus nerve branches reach both the ears and the heart via the sinoatrial node, there is a direct relationship. Stimulating the vagus nerve connection in the ear can regulate and slow rapid or irregular heartbeats. Ear acupuncture is also cited as effective in reducing anxiety and depression, both of which are affected by vagal tone.

There are reportedly 200 acupuncture points in the ear, and the World Health Organization recognizes 39 points as significant. Among the 10 deemed most important, there are specific points to engage the vagus nerve. Theoretically, inflammatory responses can be reduced by stimulating the vagus nerve to slow aggressive white blood cell production and activity, since these cells increase inflammation by releasing cytokine chemicals. (Roizen, M., 2019).

Another respected source, the National Institutes of Health, indicates that the practice of auricular acupuncture to achieve vagal tone may be able to inhibit the progression of neurodegenerative diseases, (NIH, 2019). Apparently Chinese medicine has used this technique for at least 2,000 years, and practitioners in Hippocrates time, c. 450 B.C., were reported to use forms of auricular acupuncture and acupressure to treat impotence and relieve leg pain. In more recent testing, a group of elite athletes experienced faster reduction of heart rate after exertion with auricular acupuncture compared to a control group without acupuncture. Medical evaluations confirm the effects of stimulating the vagus nerve to achieve vagal tone of the vagus nerve branch reaching the myocardium or the heart muscle.

Diaphragmatic Exercises

Diaphragmatic breathing exercises have been included in the discussion of Yoga and meditation as an essential component of the stimulation of vagal tone and initiation of the relaxation response of the parasympathetic nervous system. This section focuses entirely on the practice and effects of diaphragmatic breathing.

In its simplest description, diaphragmatic breathing is slow, rhythmic, abdominal breathing. Most people think their chest should expand on an inhale, but in fact, the abdominal area is to be pushed forward by the diaphragm, the large muscle at the bottom of the chest cavity. It is this muscle's expansions and contractions that causes the lungs to function.

In 2010, an international study reported in the *Journal of Alternative and Complementary Medicine* confirmed with empirical proof the effectiveness of diaphragmatic deep, deliberate breathing in slowing heart rate and respiration. There was also improvement in heart rate variability (HRV), which is a measure of variation within the beat-to-beat spacing, or intervals. Persons with higher HRV, mediated by good vagal tone, were found to have lower levels of stress and improved cognitive functioning, as well as overall healthier physical condition.

How to achieve vagal tone and initiate the calming, relaxation and homeostasis of the parasympathetic nervous system through diaphragmatic breathing:

Most recommended techniques simply advocate consciously breathing in deeply and slowly, extending the diaphragm muscle outward, pausing, then exhaling with equal force as slowly as the inhale, pulling in the diaphragm toward the spine. Pause for a moment and then repeat. Be conscious of these deep breaths. Continue the routine for several minutes if circumstances permit. If rushed, a few conscious, deep diaphragmatic breaths will ease at least some of the tension almost immediately

Variations include breathing only through the mouth, nasal-only breathing, or the combination. Some forms of Yoga advocate closing one nostril to make the airflow slower. Any form of deep inhalation and exhalation that is comfortable should suffice.

Carotid Sinus Massage

Massage is an easy, accessible way to stimulate the vagus nerve, increasing vagal tone and activating the calm of the parasympathetic nervous system. In its descent from the brainstem, the vagus nerve passes down the right and left sides of the neck, through the carotid sinus cavity. Being close to the skin, the vagus nerve can be stimulated by massage. This is an easy way to activate the vagus nerve, non-invasively and without electrical stimulation.

Place your fingers just above the clavicle at the base of your neck and begin to gently massage upward, to the sides. You will feel the large, firm carotid artery. The vagus nerve passes through the carotid sinus on both sides of the neck. Gently massage the front surface areas of the left and right carotid arteries to make contact with, and tone, the vagus nerve. A massage oil with essential oils can increase the relaxation sensation, but is not necessary. Continue to massage both the right and left sides of the neck for three to five minutes for fullest effect.

The massage increases vagal tone, resulting in activation of the calming parasympathetic responses. In cases of atrial fibrillation, a racing heartbeat may be slowed by this neck massage technique. In this case, the vagus nerve will send signals to the sinoatrial node of the heart, where heartbeat is regulated.

An extension of this massage can be achieved by reaching upward from the neck and gently massaging the earlobes. As noted above, the ears are reached by the vagus nerve, and earlobes are an accessible point of contact.

Vocal Cord and Facial Stimulation

The vagus nerve reaches and activates the head region, including the facial muscles, eyes, ears and the larynx and vocal cords. It is possible to achieve vagal tone, especially to relieve stress, by actively exercising the vocal cords.

Effective techniques include singing and gargling. The activity should be maintained at a moderate volume for up to 10 minutes. This technique is less immediately responsive than the diaphragmatic, auricular and carotid actions described above.

As noted above, regarding types of meditation, the vocal cords are engaged during mantra meditation via the issuance of a humming sound. In principle, this can enhance the relaxation response of meditative concentration. You may feel the vibration in the throat, as the vocal cords are vibrated by the low, base-level mantra chant. This adds a physical enhancement to vagal toning, which may increase its efficiency in activating the parasympathetic calming response.

Facial stimulation of vagal toning may be achieved easily and rapidly by application of cold. Specifically, immersing the face in very cold water for 20 to 30 seconds can have an immediate calm, relaxing effect. If this is too abrupt or shocking, application of a washcloth or towel soaked in ice water may trigger the same effect. Ideally, the facial immersion or washcloth application is repeated for a total of three cycles of 20 to 30 seconds.

In a related, recent innovation, tests are being conducted during which a person is enclosed in a chamber, and the air temperature is lowered to an extremely low level, well below zero, for one-two minutes. Those who have experienced this experimental treatment claim relaxation and at-peace benefits, but physiological tests suggest claval that these benefits are due to vagal toning, which can be achieved by less extreme, less uncomfortable means.

Chapter 5: How Trauma Affects the Nervous System

Recalling the earlier descriptions, the nervous system is actually comprised of two separate systems. One is the central nervous system (CNS), controlling the brain and the spinal cord, it's called *central* because it is the overriding controller of all bodily functions. The other is the *peripheral* nervous system (PNS), which is made of the many nerves that lead from the spinal cord and radiate throughout our bodies, extending to every extremity. Combined, these systems influence every organ and muscle, and all voluntarily and involuntarily actions, including reflexes, reactions, and responses.

Recall also that the nerves or neurons that form the substance and networks of the nervous system in the brain, the spinal cord and throughout the body, number in the billions, and form trillions of connections and interactions. Impulses that travel through and remain within the nervous system are electrical in nature, but with chemical reactions forwarding signals from one nerve to another. These are called synaptic connections.

Nervous System Overload

Considering the awesome number of nerves and their connections and interactions, it should come as no surprise that this complex nervous system can suffer from disruptions. Apart from a range of disruptive and debilitating physical conditions, including degenerative diseases, vascular and structural disorders, and infections, the nervous system can be overloaded, unable to function at full capacity and effectiveness.

When the nervous system becomes overloaded from stress, depression, anxiety, or from any number of factors that push the nervous system past a certain point, functionality becomes reduced and the body enters a state of fatigue. In these cases of system overload, fatigue cannot be traced to muscular deficiencies nor to depletion of glucose and other energy stores, and it is not resolved by rest. Something else is at work, affecting electro-chemical connectivity and reactivity.

There are many symptoms of nervous system overload, including persistent headaches, sudden onset headaches, and headaches that changes or may feel different, tingling sensations, and feelings of muscular weakness that cannot be attributed to physical exertion. Other symptoms may be loss of sight or experiencing double vision, loss of memory and impaired thinking ability, and a lack of coordination. In extreme situations, a person may suffer what is commonly called a nervous breakdown, left in a state of near immobility and unable to think clearly or rationally.

Central nervous system fatigue can occur as the result of excessive physical exercise. The molecular composition of the neuro-transmitting chemicals that connect nerves to each other changes, slowing or altering the synaptic connections between the nerves, and this may slow or even block the neural connections between muscles and organs, and the brain and spinal cord.

Overcoming nervous system fatigue:

Several commonly available substances have been proven under medical testing to have a positive effect on preventing or reversing nervous system fatigue: coffee, carbohydrates, and amphetamine.

> The caffeine in coffee acts to suppress production of adenosine, a chemical neurotransmitter that can induce fatigue. Athletes are cautioned to

ingest coffee in moderation for optimal effect and for safety.

Ingestion of a combination of a readily-available carbohydrate. e.g., sugar, and an electrolyte, appears to increase plasma levels of glucose, a primary energy source. However, there is no increase noted of glycogen within muscles.

Taking amphetamines is commonly used by performance athletes to prevent or reverse central nervous system fatigue. It is known to alter the production of the neurotransmitters dopamine and norepinephrine, which play a role in inducing nervous system fatigue.

Chronic fatigue syndrome is characterized by physical fatigue that is not caused by peripheral muscle fatigue. As with nervous system fatigue, it is persistent, not caused by exercise and not alleviated by rest. Subject suffering from this condition were unable to compete exercise tests, despite no obvious physical reason for their limitations.

The main cause of chronic fatigue syndrome has been traced to the central nervous system, where certain defects are slowing or otherwise altering neural connections. In other words, the muscles and organs of the body are intact and potentially fully functional, and it is the connecting electrical wires that are at fault. Precise medical treatments are still being explored, while the fatigue-relieving self-applications, like caffeine from coffee, carbohydrates with electrolytes, and amphetamines, are found to help along with motivational counseling.

Daily stress from a multitude of responsibilities and interests can lead to nervous system overload without any physical causes. Contemporary lifestyles include long work and study hours, insufficient time off, or a sense of guilt when time off is taken, an all-encompassing range of video and online distractions, plus personal and family responsibilities. In response, sleep deprivation may occur, multitasking becomes the standard work-and-activity management tool, and stress builds until it overwhelms the nervous system.

Situations that have reached nervous system overload may evoke the proverbial cup runneth over. One result may be activation of the sympathetic nervous system's fight or flight response, with surges of energy that are quickly depleted resulting in chronic fatigue, shortness of temper and irritability, and even in less extreme situations, a reduction in effectiveness and the absence of the satisfaction of fulfillment.

More extreme trauma may create nervous system overload. Involvement in an accident or a physical injury, for example, or a strong emotional shock, such as unexpectedly being fired or the ending of a close personal relationship. In these cases, as with the other situations, the parasympathetic response may cause overreaction. Consider the behavioral eccentricities of some extreme cases, including violent behavior.

How to respond to emotional, non-physical causes of nervous system overload?

The practical applications of the Polyvagal Theory may be successfully applied to diminish and perhaps eradicate the effects of nervous system overload. These include Yoga, meditation and managed breathing exercises, and the other procedures such as neck massage and cold facial treatments. The objective is to achieve vagal tone and activate the relaxation response and calming effects of the ventral vagal parasympathetic nervous system.

Degrees Of Stress

Stress. A word frequently used to describe everything from a simple sense of urgency to the extreme disabling conditions associated with long-term, overwhelming stress. We tend to think of stress as a modern affliction, resulting from the overload of obligations, media inputs, personal interactions both personal and professional, and everything brought to us online, such as social media, email, text messaging, streaming video, customized music playlists, news alerts and stories, magazine articles, eBooks and audiobooks, sports programs we can carry with us, and more.

Actually, people have always experienced stress due to any variety of outside influences, as well as internal sources, stemming from imagined or real concerns, fears, and worries. But, yes, today's environment seems to be overflowing with more than we can handle, leading to an almost infinite number of stimuli and distractions, engagements and commitments.

Consider the increased levels of our exposure to programming content and advertising. From historical print-only to radio, then the emergence and dominance of television. It was first local, then networked, national broadcasts, followed by cable broadcasts with huge numbers of watchable channels covering every interest. Add to this, today's technology-enabled out-of-home media that presents us with product information and advertising at the point-of-sale in supermarkets, drug stores and mass merchandisers, and even at the service stations to entertain and inform us while we're pumping gas. Of course, outdoor ads like billboards have been around for over a century, but today, a walk through a shopping mall will expose us to dozens, or sometimes hundreds, of electronic billboards, some of which are sophisticated enough to use facial recognition software to personalize ads as we approach. All-in-all, it's a lot to absorb, manage and digest.

Apart from this vast multitude of inputs, consider the many distractions and stimuli that the pressures society imposes upon us, plus the pressures we impose upon ourselves. Work, education, sports, even vacations can incite stress. How am I doing compared to my peers? Will I qualify for whatever you are aiming for? Do my parents love me? Does my spouse love me? Most of us manage to find a good number of concerns to think about, dwell on and worry about, and if we don't manage these concerns, they can become forms of stress in our lives.

In reality, stress is an emotional condition that psychologists classify into three degrees of seriousness on both emotional and physical levels:

1. Acute stress:

The most common form of stress, is frequently the result of intense, recent psychological pressures. These can be in the recent past, such as a scary, near-miss auto collision or concern over a just-ended difficult meeting, a tough exam or an interview just completed, and the outcome of which is of concern. Or, acute stress may be triggered by anticipation of a future event, like a sports challenge, a pending confrontation with a boss or a spouse, a scheduled operation, or dental procedure. Because acute stress tends to be short-term, even if extreme, it does not tend to cause serious side-effects or injury.

Emotional reactions to acute stress commonly include tension, anger and irritability, anxiety, and depression. Physical reactions may include stomach upset, loss of appetite, diarrhea or constipation, acid indigestion or chronic acid reflux (GERD). Also included can be tension headaches and migraine headaches, muscular tension, elevated heartbeat or heart palpitations, shortness of breath and lightheadedness or dizziness.

Many of these are symptoms of a defensive sympathetic nervous system response and may be alleviated by toning the vagus nerve and inducing the relaxation and social engagement responses of the parasympathetic nervous system. Meditation, Yoga stretches and poses, and managed deep breathing exercises can alleviate the feelings and symptoms of acute stress.

2. Episodic acute stress:

While ordinary acute stress is short-term and relatively harmless, when the patterns of stress are not occasional, or are due to multiple causes, a more extreme form of stress, called episodic acute stress, occurs. Episodic acute stress is longer-term, involves a multitude of stress-inducing factors, and is harder to manage and bring under control. It is often the result of an overloaded, excessively ambitious lifestyle, for example, the "Type A" personality defined by cardiologists (Friedman, M., and Rosenman. R., 1976). These are people who are excessively driven by often unrealistic goals, time urgency, and high levels of expected accomplishments. They tend to be aggressively ambitious, irritable, and impatient. Many are driven by a deep-seated insecurity and are outwardly overambitious in compensation to their insecurity.

Another type of person who suffers from episodic acute stress is the persistently worrying personality, like persons who have a pessimistic outlook on life, expect the worst, anticipate nothing working out. They are the antithesis of optimists. Unlike others who experience stress, these worry warts tend to be less angry and irritable, but are more likely to experience continuing bouts of depression or anxiety.

Emotional and physical symptoms of episodic acute stress are similar to those of acute stress, but being longer-term in nature, can induce more persistent and severe headaches, anxiety, depression, chest pain, and in the case of the Type A personalities, a greater propensity to develop heart disease.

Treatment of episodic acute stress is more difficult on several levels as the conditions it causes are more severe and chronic. Also, the personalities of these types of people make them resistant to behavioral and attitudinal changes. Type A personalities are convinced that their driven behavior is the right and only way, and others are lazy slackers. They believe nothing will get done without their excessive presence, pressure and micromanagement. Those who worry excessively are convinced that their concerns are realistic and necessary, believing that there is nothing positive to anticipate. In consequence, treatment to modify these attitudinal and behavioral extremes is complex, and often needs to be at the professional level.

3. Chronic stress:

This is the long-term, potentially damaging stress of hopelessness. People experiencing this form of stress experience lives of continuing frustration, and see no way out and no ending to the trouble. Examples include women in loveless marriages and married men who realize they love someone else, who in turn, is also married and unavailable to them. It includes people in dead-end jobs they despise and who believe their lack of education, or their ethnicity, or their handicap or other physical characteristics will keep them from advancing or going elsewhere, and somewhere better. It can embrace people living in areas of violent conflict, whether in the shifting political turmoil in the Middle East, the poverty of certain African countries, or the gang-incited violence of some Central American countries or American inner cities. The common characteristics of chronic stress are senses of entrapment, a bleak future, and no way out.

The consequences of chronic stress can be severe, including deep depression, a permanent state of anxiety, and unending hopelessness. Given its continuing nature, people with chronic stress may be unaware of its existence and its hold over them. They surrender to it, accepting it as normal, the way life is and was meant to be. Over time, chronic stress can leave the individual vulnerable to a range of diseases like mental illness, constant fatigue and muscular weakness, heart disease and a propensity to be suicidal. Chronic stress sufferers may be irritable, resentful of others who are not experiencing the same misery, and they may be subject to violent outbursts. Cases of someone *going postal* usually involve a person whose work situation was depressing, hopeless, and frustrating. These conditions can lead to reduced effectiveness on the job, thus accelerating a downward spiral, and leading to reprimands or termination. These actions, in turn, can become pivotal moments that trigger violent actions, often directed toward fellow workers.

Professional help is generally required to effectively pull a chronic stress sufferer back to a sense of normalcy, especially when they consider their unhappy condition to be inevitable. Treatment tends to require long term psychological counseling and therapy. The symptoms may have begun early in life and are deep seated. In addition, medical treatment may be required when the chronic stress has provoked long-term weakness, and muscular atrophy, and the onset of cardiovascular disease.

Panic And Hyperactivity

People who are subject to hyperactive behavior are most often diagnosed as having Attention Deficit Hyperactivity Disorder (ADHD). There is a tendency to experience bouts of panic, both short-term and longer-term, which affects 50% of adults and 35% of children who have been diagnosed with ADHD. The panic induces anxiety, and even at low levels it can exacerbate the customary ADHD behavioral effects: hyperactivity, nervousness, fidgeting and inability to sit still, as well as challenges to concentration and the completion of projects. In effect, hyperactivity and panic-induced anxiety tend to coexist. For example, an inability to focus or concentrate for a sustained period is characteristic of panic-induced anxiety and hyperactivity.

But there are symptoms that define panic-induced anxiety that exceed the normal bounds of ADHD. These include long term or chronic feelings of nervousness, a tendency to continuously worried, irritability, insomnia or other sleep disorders, frequent headaches and backaches not caused by physical strains, and fears of the unexpected and unknown.

Treatment to reduce or relieve panic-induced anxiety may be sought by discussing the symptoms with medical professionals. Counseling may advise self-evaluation to identify the triggers that create the panic, leading to recommended behavior modification. For example, if the making of deadlines is a source of panic reactions, the person may be encouraged to plan more carefully to allow more time to accomplish responsibilities.

Many of the symptoms of panic and anxiety may be due to activation of the sympathetic nervous system response, the call-to-action mechanism with its characteristic elevated heart and respiratory levels, and shutting down of digestion. In less extreme situations, these defensive reactions may be alleviated by countering the sympathetic nervous system response, bringing the ventral vagal response into play, as discussed above, with Yoga poses and stretches, meditation and achievement of mindfulness, and deep, forceful and conscious breathing.

Separately, it is possible that medications prescribed for ADHD can cause panic or anxiety tendencies. Physicians can evaluate the potential side effects of medications, and advise changing to those less likely to cause panic reactions.

Chapter 6: The Polyvagal Theory And PTSD

PTSD, or post-traumatic stress disorder, has gained considerable attention in recent years due to its occurrence among military veterans, especially those returning from the long, ongoing conflicts in the Middle East. These traumatized individuals may have experienced severe physical injuries, but in many cases, however, their injuries are psychological, resulting from their overwhelming reactions to their battlefield experiences. In earlier wars, mentally traumatized veterans were said to be suffering from shell shock, the result of seeing and feeling the consequences of war. We now recognize this condition as PTSD.

Typical symptoms of PTSD include flashbacks of the traumatic event or the inability to stop thinking about it obsessively, anxiety, depression, sleeplessness and recurring nightmares. Beyond the discomforts of experiencing PTSD, it is now known that it can lead to suicidal thoughts and suicidal behavior. In many cases, PTSD can lead to continuing deep depression and anxiety, as well as eating disorders, and substance abuse, notably drugs and alcohol.

Apart from veterans, people in all walks of life may have had terrifying, traumatic experiences, either themselves or as witnesses, that trigger PTSD, like an automobile accident, sexual or other physical assault, a serious fall at home, or loss of a loved one. Any of these extremely distressing experiences may initiate the PTSD response. Previously, victims of PTSD may have been told to shape up or get over it, but today, PTSD is a recognized, serious psychological condition requiring professional assistance to resolve. It may affect children as well as adults.

Based on the Polyvagal Theory, it is now believed by many psychologists that PTSD has its roots in the dorsal vagal response of the parasympathetic nervous system. This is the primitive freezing, or shutting down mechanism that is triggered when the person or animal faces an insurmountable or overwhelming immediate threat. When this dorsal vagal response is initiated, it can cause immobility, speechlessness, fainting and even severe shock. PTSD appears to be an ongoing form of dorsal vagal reaction.

Before reviewing the Polyvagal Theory's potential treatments for overcoming dorsal vagal-caused PTSD, an understanding of the human brain's evolution and functions is presented for perspective.

The Three-Part Brain

The human brain, with its complexity of 100 billion or so neurons and perhaps 100 trillion neural connections, is generally known to be organized into two hemispheres, the left, recognized for controlling rational, logical, organizational thoughts, and the right, associated with creative, imaginative and unstructured thinking. We also know that the functioning nervous system is comprised of the brain, spinal cord, and between them, the brainstem.

>The brain is where all the conscious and unconscious action takes place, from managing our cardiovascular, respiratory and digestive functions to feelings, senses and sensations, and embracing all thought, memory and decision-making.

>The spinal cord is the central cable that receives all nerve impulses from the extremities and

forwards these impulses to the brain, and returns the brain's reactions to the impulses with the appropriate reaction.

The brainstem is where 10 of the 12 cranial nerves originate and extend to the organs and other key areas, including number 10, the longest, most diverse neuron, the vagus nerve.

But we know today that the evolution of the human brain has been built upon a sequential three-part structure, beginning with the earliest, most primitive part, called the reptilian brain, then continuing to evolve an early old or paleomammalian brain, and concluding with a more sophisticated new or neo-mammalian brain. This concept of a three-part evolution-driven brain structure was identified first in the 1960's by a neuroscientist, Dr. Paul MaLean, who called it the triune brain, and postulated that these three parts of the brain still struggle to coexist. Each part has specific functions to perform:

> The early, *reptilian* brain, is responsible for basic, involuntary reflex actions, including reproduction urges, arousal to a range of stimuli and maintaining a balanced, normal state, or homeostasis. It can be considered a fundamental

survival mechanism. One of its continuing characteristics is compulsiveness.

The old-mammalian, or *paleomammalian* brain, is positioned to surround the reptilian brain, it manages emotions, learning and memory functions. It enabled early mammals to remember and act upon favorable and unfavorable experiences, for example.

The new-mammalian, or *neo-mammalian* brain is responsible for conscious thought and self-awareness, and is positioned atop the two early brain parts. All of our reasoning, decision-making and rationalizations occur here.

But one may ask if we really evolved from reptiles? The concept of our brains evolving from reptiles comes as a surprise. We understand that we evolved from mammals, since we ourselves are mammals. Okay, but reptiles? Over the long course of evolution, the earliest mammals evolved from, yes, reptiles, and not from the dinosaurs that became extinct 66 million years ago, or the dinosaurs that grew feathers and evolved into birds. Our reptilian ancestors were small, and obviously smarter than the large dinosaurs, which gave them an edge in natural selection. They had strong survival skills built into their small but highly functional reptilian brains, and some of these hardy reptiles evolved into small mammals. In their turn, these early mammals evolved more complex brains, the paleomammalian brain, with its added values of learning, memory and emotion. Still later, as mammals further evolved as primates, the third neo-mammalian brain component developed, giving Homo Sapiens the ability to think consciously and with increasing complexity.

The three parts of our current triune brain correspond, approximately, to the brainstem and cerebellum (reptilian), limbic brain, which includes the hippocampus, amygdala, and hypothalamus (paleomammalian) and the neocortex (neo-mammalian). Because the reptilian-originated brainstem reacts completely unconsciously and immediately for survival, historically, it tends to dominate in many situations, when the brain perceives a danger or other need for prompt action. The conflict between the purely instinctive reptilian brain and the two more advanced components is considered by some to be represented by Freud's ongoing battles between the conscious and the subconscious.

When the other dimensions and aspects of the brain are considered along with the three triune sections, the complexity of brain functioning begins to become clear. These aspects include the two-hemisphere structure, vertical networks connecting the layers and departments of the brain, and a near infinite number of interacting neurons, as well as variations in brain structure due to gender, genetic and environmental influences.

In recent times, the precise sequential evolution and functioning of the triune brain, and its exclusivity among humans have been questioned by some animal behaviorists, since complex brains have developed among non-mammal species, including certain birds. Also, new studies demonstrate that in humans, the prefrontal cortex performs complex functions that are apart from the functions of the neocortex.

Post-Traumatic Brain Reeducation

Separate from the psychological disorders associated with PTSD, there are physical brain injuries resulting in serious trauma. About 10 million people worldwide suffer traumatic brain injury (TBI) each year, and many cases are fatal, and most who survive the injury experience some degree of cognitive impairment. These trauma may occur in any number of circumstances, including vehicular accidents, sports injuries, falls inside and outside the home, acts of conflict or violence, even being struck by falling objects.

There are a range of treatments to reverse the impairment, and the type and duration of treatment depends on the type and severity of the trauma. Generally, a multidisciplinary set of treatments is required, involving the psychiatric and neurologic medical practices, as well as pharmacotherapy.

Classifying TBI as mild, moderate or severe depends on several key factors: Degree of post-traumatic consciousness, duration of the coma, if experienced by the patient, and the degree and duration of post-traumatic amnesia. Generally, TBI patients whose symptoms continue for one month or more are classified as either moderate or severe, and whose full recovery make takes years, while those showing marked improvement within a few weeks are considered to be mild cases and often return to full cognitive function within two months.

There are a number of impairments to the cognitive functions following TBI. These are the most commonly treated:

- Decreased ability to concentrate
- Impaired attentiveness
- Reduced visual spatial cognizance
- Tendency to be easily distracted
- Memory lapses and impairments
- Loss of executive ability (decision-making)
- Disrupted communications skills
- Judgmental lapses and dysfunctions

Reeducation of TBI patients begins with assessments based on standardized testing protocols, including visual and auditory attentiveness, visual and verbal measurements, language comprehension and understanding, executive function (decisiveness), overall mental and intellectual function and motor function.

Post-traumatic brain reeducation is undertaken primarily through cognitive rehabilitation, which works to increase the injured person's abilities in the processing and interpretation of information, and the overall performance of mental functions. Cognitive rehabilitation is mostly effective in mild or moderate levels of TBI and with persons who have a high level of motivation to succeed in the recovery. The multidisciplinary group that collaborates on brain reeducational therapy may include doctors, speech and language specialists, physical and occupational therapists, among others. However, it is recognized that each patient's treatment will be unique, prescribed and tailored to each individual, based on the specific injuries suffered and resultant trauma.

One important approach that has wide application is attention process training (ATP), which is based on mental skills training, gradually increasing the complexity of the exercises, from simple initially, and subsequently increasing in complexity, forcing the brain to retrain itself. The exercises include selective attention, focused attentiveness, alternating attention, divided attentiveness and sustained attentiveness.

The Parasympathetic Recovery

The Polyvagal Theory links PTSD to one dimension of the parasympathetic nervous system (PNS), the early-evolved *dorsal vagal* freeze survival mechanism. The dorsal vagal mechanism may protect an animal by allowing it to play dead until the coast is clear, but in a human being, it can lead to inaction, inability to think or speak, or worse, passing out or fainting, shock or even cardiac arrest. With the linking of PTSD to the dorsal vagal mechanism, a previously unrecognized cause may now be open to evaluation and potentially, to alleviate the symptoms of PTSD.

Specifically, the other, more recently evolved PNS response, the calming, relaxing, socially engaging *ventral vagal* response may be applied to reduce the emotional and physical symptoms of PTSD. Now the methods used to achieve vagal tone and lower heart rates and breathing rates, reactivate the digestive system and induce an all-encompassing state of calm and relaxation may be applied by the individual, easily, every day. These methods, as we've discussed, include meditation, Yoga stretches and poses, and managed, deep and conscious breathing. The practice of deep, slow breathing, with forceful extension of the diaphragm to tone the vagus nerve, is applicable as part of meditation or Yoga, or simply done without other techniques.

It can also include auricular and facial massage, massage of the vagus nerve as it passes next to the right and left carotid artery in the neck, and cold facial therapy. The practice of mindfulness, or being in the moment, in which all outside thoughts are prevented from intruding, can also be beneficial, as the person concentrates on every external sound, every feeling, every awareness of things in the environment. Vocal stimulation of the vagus nerve can be done easily by singing, gargling, or reciting a mantra while performing mantra and transcendental meditation.

Another application of Polyvagal Theory to treating PTSD is for the individual to recognize that the symptoms of PTSD are biological in nature, caused by the body's primitive instincts and reflexes to protect itself, and that the body can be taught to relax, get over it, rejoin and socially engage with those who are living active, normal lifestyles. This is called somatic awareness, and it trains the individual to become aware of basic bodily functions like heart rate and breathing, and to consciously try to slow them down. The deep breathing exercises may be helpful in achieving a sense of bodily control.

The reduction or elimination of PTSD symptoms can further be achieved by practicing a series of mental exercises called attentional control, a conscious effort to recognize the cues that may trigger PTSD reactions, and gently but firmly cancel them out by acknowledging that there is no danger, nothing to fear, and all is well. This form of body awareness is called cognitive behavior therapy (CBT), and it encourages the individual to be aware that an unneeded fight or flight response is continuing and can be shut down by conscious thought, replacing disturbing thoughts and memories with relaxing, peaceful thoughts. Over time and with practice, the replacement of bad thoughts with positive ones will make the cooling down of the dorsal vagal action-orientation easier.

Reading Body Language

Body language has long been associated with a few popular positions and movements that are believed to be subconscious cues as to a person's true meaning or intentions. For example, having one's arms crossed signals a negative interest in what is being said, or a hand over one's mouth while speaking may be a sign of a lie being told. Unconsciously nodding one's head indicates agreement, a handshake suggests type of character, depending on whether it is firm or weak, and if eye contact is maintained or not. In reality, most of these body language cues are anecdotal and may have some basis, or they may not.

But Polyvagal Theory has shed a new light on body language, on multiple levels, by revealing one's interest in a social engagement, for example, or sending a signal that can trigger social engagement or other interaction in the second person, who may, in turn, respond with their own body language subconsciously. The use of facial expressions to elicit various types of responses is being used to communicate and engage with autistic children, in testimony to the effectiveness of this approach. (Communicating and engaging with autistic children is covered in a subsequent chapter.)

Do the popular body language signals really mean anything, or are they, as implied above, merely anecdotal, believed and circulated but without substantiation? A study conducted by UCLA found that only 7% of what is said is actually believed or acknowledged, based only on the words spoken. The tonality of the speaker's voice accounts for 38% of communications, leaving 50% of communications being based on body language, gestures and expressions.

Resistance to what is being said or shown is frequently shown by crossed arms and crossed legs.

A smile is not sincere when it is limited to the mouth, whereas a sincere smile engages more of the face, including crinkling the eyes.

Mirroring or imitating your own body positions is a sign that the other person is in agreement with what you are saying or proposing.

Power positions radiate a sense of command or control. A person who assumes control will tend to stand upright, extend arms and otherwise occupy more space in a room. This type of person is encouraging interaction or possibly engagement.

Eye contact is not always synonymous with engagement or interest because extended or prolonged eye contact may be forced or deliberate, suggesting the person is hiding a true intention.

Discomfort or surprise may cause raised eyebrows. Conversely, a truly interested person will not tend to raise their eyebrows when spoken to, except to acknowledge an exceptionally unusual remark.

Nodding is positive, except when it's exaggerated because too much nodding suggests discomfort with what is being said.

Tension signals stress. A furrowed brow, tightened neck muscles or a clenched jaw may be signs that what is being said is making the person uncomfortable.

Are these findings valid? Probably to some degree, but it's important to realize that the subject of body language has been widely discussed and debated for decades. As a result, many people you may be speaking with, or meet in an interview, may be consciously nodding or smiling or firmly shaking your hand, deliberately trying to make a good impression. You, in turn, might consider your own body language, and try not to send the wrong message.

Chapter 7: The Polyvagal Theory And Emotional Stress

Among the more profound conclusions emerging from Dr. Porges' Polyvagal Theory is the linking of the emotional and physical responses we are subject to. Emotional reactions can trigger not one but two physical responses: the well-known defensive call to action of the sympathetic nervous system, and the more primal dorsal vagal response that can freeze and immobilize a person. Physical actions, conversely, like Yoga, meditation, managed breathing and massages can tone the vagus nerve, triggering the calming, relaxing emotions of the parasympathetic nervous system (also called the ventral vagal response), and its enablement of social engagement.

Normal or Interactive State

The concept of social interaction or social engagement is far more complex than children playing with each other, or adults attending a reception or a party, or joining their neighbors at a summer evening barbecue. Social interaction is actually an advanced state in the hierarchy of defense mechanisms that have evolved to protect mammals, notably primates, from the risks and dangers of the unknown or unfamiliar. These protective actions begin with primitive rejective reactions and culminate with interactions and engagements that provide a sense of community, where there is mutual recognition, support and protection.

The Polyvagal Theory introduced the concept of *neuroception*, whose purpose is to analyze and interpret environmental factors, and then to initiate either defensive reactions, or to stimulate the onset of the calming reaction. A common example is the function of neural circuits that enable a child to smile and respond positively when recognizing someone familiar, but to hesitate or flee from an unknown stranger.

In this scenario, neuroception is both encouraging the safety and positive benefits of social interaction and providing defense mechanisms for protection. This is due to mammals, especially primates, having evolved and developed structures within the central nervous system that manage and regulate responses to both social-inducing and danger-threatening situations. According to the Polyvagal Theory, these structures came into being sequentially, so while the first may appear as an obvious defense mechanism, in reality all three are unconscious reactions that are purely protective in their nature.

The first protective reaction to form is immobilization, which is the hit-the-brakes response to threats posed by the unknown and, by definition, the untrusted. This may be when encountering a stranger, being afraid to enter a dark room or a long alley, or be hesitant to walk near a group of noisy teenagers, since any person, or animal, or situation that is unfamiliar can be interpreted as threatening. This response—immobilization—causes an infant to cry, and a child or an adult to stop, to turn away, to reject the other's advances, and to seek shelter. In earlier times, it may have been more likely to be a true physical threat than we might typically encounter today, but the reaction is the same among infants, children, and adults. This is true, even if the situations that trigger immobilization are different for each. An infant may hesitate or cry simply because a stranger is near, a child may stop before entering a new school, an adult may become immobile when attempting to cross a busy intersection, to enter a smoke-filled room, or to advance toward a contentious crowd.

The second defense, less extreme, is mobilization, which subconsciously puts us partially at ease because of certain reassuring factors. These may include recognizing a person we had met previously, or it could be that we are comfortable with their manner of dress. This is why we like to dress up to make good impressions among strangers or those who may be in a position to evaluate us. Other forms of reassurance may be a uniform worn by a policeman or a postman, or even a FedEx or UPS delivery person, as these uniforms actually serve a purpose of facilitating recognition and familiarity. So even if the person wearing the uniform is unknown, the appearance can provide sufficient reassurance for a person to feel at ease and acknowledge the other person by exchanging a greeting, asking for help or directions, or opening a door. Other forms of reassurance to encourage and enable mobilization can be a stranger's facial expression or tone of voice, and even a handshake and warm eye-contact can be sufficient for a bond of trust to form.

The third defensive mechanism is engagement, or social interaction and attachment-forming, which are more than simple social courtesies. The successful evolution of a number of primate species, Homo Sapiens certainly included, depended upon acceptance by, and cooperation with others. During these earlier, pre-agricultural times, when life depended on effective hunting and gathering, single individuals and small independent families could not compete with larger groups. These early organizations tended to have sufficient manpower to diversify roles, allow specializations to develop and be able to divide labor based on abilities and aptitudes. Acceptance into a larger group become a necessity for survival. Today, the need to be accepted by others is not necessarily a sign of insecurity, rather it's a holdover of a survival mechanism. Terms like social engagement and social interaction now encompass neighborhood associations, and affiliations and clubs based on any number of common interests. Religious affiliations are a huge, extensive example of people with common interests coming together physically and with regularity. Schools and universities, by their nature, are places where social interaction is diverse and extensive—consider the range of teams, social organizations and clubs based on activities and interests in any academic environment. Joining, participating to some degree, engaging with others in your group, yes, this may be for enjoyment or learning, but it's also for the security of group acceptance and for some, perhaps to a subconscious degree, a sense of protection.

Remember that these three defensive, protective mechanisms are naturally evolved, continuing emotional and physical responses, based on neuroception, the analyses and response to potential threats. They are subconscious, reflexive, and operating without our knowledge.

Emotional Stress-Induced Sympathetic Response

Research has verified and quantified the degree of physical responses that have been triggered by emotionally induced reactions to stress. Over 900 subjects participated in a "Midlife in the United States" study by Columbia University that was reported by Elsevier in 2011. Exposure to stress created the classic signs of a sympathetic response, notably elevated heart rates and respiratory rates. Depending on each individual's personal traits, reactions were classified as either mild or severe. Those with mild sympathetic responses rapidly returned to normal with high levels of vagal toning, while those with severe reactions to stress required a longer recovery time.

In another smaller-scale study (N=70 participants), conducted by the University of Bergamo, Italy, vagal tone and reactions to stress were evaluated using respiratory sinus arrhythmia (RSA), which is a measure of how the heart rate varies during each inhale and exhale, also known as heart rate variability (HRV). When stress was induced during cognitive tests that required high levels of focused concentration, it was found that those with pre-stress high levels of vagal tone responded at lower levels of sympathetic nervous system defense. Those with lower vagal tone before the stressful exposure had more intensive and longer lasting sympathetic responses.

Conclusions from these and other studies verify the proactive role of the vagus nerve in reducing the degree of reaction to mild-to-moderate stress. Some interpretations of these findings are that over time, the vagus nerve has evolved a protective mechanism to prevent stress-induced overreactions, preventing the sympathetic nervous system from initiating the full fight or flight defense response. Individuals who have developed this mechanism are found to have higher levels of baseline vagal tone under normal, stress-free circumstances.

Taken further, it has been found that those with higher levels of baseline vagal tone are generally more relaxed, have positive emotions and are more open to social engagement than those with lower baseline vagal tone. It is believed that evolution's process of natural selection has led to better vagal tone, thus protecting the individual from the shocks of the sympathetic nervous system's defensive calls-to-action, as well as from the parasympathetic nervous system's extreme freezing, immobilization, fainting and shock reactions to the vagus nerve's dorsal vagal responses.

These two parasympathetic nervous system responses are diametrically opposed, as discussed in the following section.

Shutdown or Calming Parasympathetic Response

By way of review, the autonomic nervous system had traditionally been thought to be comprised of two opposing systems that operate without our knowledge or assistance: the sympathetic and parasympathetic nervous systems.

The parasympathetic nervous system has historically been considered to be the second of the two autonomic nervous system components to evolve. The first to come into being, the sympathetic nervous system, is responsible for fast, intensive reactions to danger and serious risks. It is believed to have come first because it is an immediate response to existential threats. The sympathetic response speeds up the cardiovascular and respiratory functions, causing elevated heart rate and shorter, faster breaths. It stimulates the release of norepinephrine (adrenaline) and causes a surge of glucose into the blood to give the cells of the muscles an extra boost of energy, while shutting down the digestive system and other metabolic activities to conserve energy. The individual is ready for peak performance, either to escape and evade danger, or confront it head-on. It's the often quoted as the fight or flight response.

Later in our evolution, it is believed, the parasympathetic nervous system came into being to bring the individual down once the threat or danger is resolved and under normal circumstances, to keep the individual in a state of normalcy or homeostasis, with a slow, regular heart rate, and normal, slower, deeper breathing, and a digestive system that continues its steady work of peristalsis—contracting and expanding rhythmically to move food through the stomach and small intestine. The parasympathetic nervous system is mediated by the vagus nerve, specifically, the frontal or ventral vagal function. In addition to keeping the body on an even keel and in a relaxed state of calm, the ventral vagal encourages social engagement, which, as we have seen above, is actually a defense mechanism, albeit a subtle one, working at a subconscious level. It manages how we meet and greet, or otherwise react to others, from immobilization during uncertainty to mobilization when concerns are alleviated, and then onto social engagement and interaction, helping us to benefit from group interaction and mutual support.

But the Polyvagal Theory has revised the thinking, so that there are now believed to be not two but three autonomic responses, the dorsal vagal or freeze response has achieved recognition, and is considered the most primitive of the three autonomic system components. Interestingly, this severe shutting down of the central nervous system is considered to be a function of the otherwise calming, stabilizing parasympathetic nervous system.

The dorsal vagal, freeze response is a reaction in extremis because it stops all action. In nature, it is believed to have begun among reptiles, and continued its presence and function as reptiles evolved into small mammals, and further continued as mammals evolved to larger, more complex primates, and eventually, to us, Homo Sapiens. The dorsal vagal response is believed to be controlled by the brainstem, which is located between the brain and spinal cord, and is believed to be the part of the nervous system that evolved first. The brainstem is where most of the 12 cranial nerves originate and spread throughout the body to the key organs. Among these essential cranial nerve, the 10th is the vagus nerve, the longest and most diverse of the 12, responsible for mediating both the calming, engaging ventral vagal response, and the freezing, shutting down dorsal vagal response.

The dorsal vagal response is initiated when a creature—or a person—is confronted with an overwhelming threat or other extremely stressful situations, and especially when there does not appear to be possible action and there is no way out. The dorsal vagal response complies by preventing movement. For a small reptile or mammal, the response may be to play dead. This is not something the animal is doing consciously, it is not thinking, "I'll just pretend I'm dead, and the fox will move on and then I can escape." The animal has involuntarily fallen into a trance-like state of immobility. It will be able to sense when the threat is still present, and later, when the fox has moved on, when the coast is clear and the animal can escape or seek shelter. The dorsal vagal response is extreme, but it has its evolutionary purpose in supporting survival. Natural selection has provided a survival option, permitting animals to live another day, and be able to reproduce equally well-equipped progeny.

A human being's dorsal vagal response may be initiated when overwhelmed with fear, when in an accident, or when subjected to physical harm, or the perceived threat of physical harm. It can happen when entering a room and suddenly feeling one is in the wrong place, or facing people not known or recognized. When the dorsal ventral response is triggered, a range of immobilization effects may occur, from simple speechless to being unable to take a step, or to more severe actions, including dizziness or lightheadedness, disorientation, fainting, and, in the extreme, entering a state of shock. In rare cases, the shock can lead to heart failure, organ failure, or cardiac arrest. In these extreme overreactions, it is due to a primal defense mechanism that has exceeded its original, protective mission.

Not all causes of a dorsal vagal response are serious, life-threatening or shocking events. A person may become lightheaded, disoriented, immobile, or may faint simply by standing up too quickly, or engaging in vigorous activity without warming up. The body reacts with a dorsal vagal response by dropping the blood pressure and slowing the heart, instead of the usual increases in heart rate and blood pressure when stressed. In these less-threatening situations, the immobility or lightheadedness passes quickly and the body returns to normal homeostasis. Generally, the calming parasympathetic nervous system is keeping watch and can recognize the need to control heartbeat and breathing rates, quickly returning them to normal.

A person who enters a mild state of dorsal vagal immobility or lightheadedness, for example, can take immediate action to initiate relief though activation of the parasympathetic nervous system. Standing in place, deep diaphragmatic breathing can be started as the advice, take a deep breath, has a biological basis. Taking up to 10 deep, slow breaths with forceful, thoughtful inhalations and exhalations, can quickly return a person to a state of calm equilibrium. If possible, sitting down with the head lowered can reduce the lightheaded sensation. Focusing on a single object while deep breathing can enhance the effectiveness of the movements. If possible, some light stretching can accompany the breathing exercises.

Chapter 8: The Healing Power of Vagal Tone

The Polyvagal Theory has effectively linked the physical and emotional. Physical actions can regulate emotional conditions, emotional activities can cause physical responses. For example, deep, forceful, diaphragmatic breathing can initiate a state of deep calm, while emotional reactions can lead to stress, triggering elevated heart rate and respiratory rates and a range of other visceral organ reactions, such as stopping digestion to conserve energy. Given the role of the vagus nerve in mediating both physical and emotional reactions, it is no surprise that the vagus nerve can be engaged to better manage our emotional sense of well-being and help alleviate physical problems.

As we have seen, under normal conditions, the calming parasympathetic nervous system is dominant, keeping the body in a state of homeostasis. In this context, vagal tone is an assessment of the body's readiness to perform certain key functions effectively. An ideal vagal tone maintains a baseline from inputs, via the vagus nerve, received from the parasympathetic nervous system. Among the most important vagal tone functions is controlling heart rate to keep it from beating too quickly. Vagal activity is key as well to controlling breathing rate, managing the rate of peristaltic contractions during digestion, and further affecting the sensitivities and inflammation of the digestive tract and functioning of the liver. Vagal tone is also a measurement of emotional stability, as emotions are at their baseline of normalcy when the dorsal vagal and ventral vagal responses are at homeostasis.

But this is not always the case, especially when emotional reaction ignite physiological responses.

Regulating Emotion

The parasympathetic nervous system follows two pathways. The better known, and far more dominant, is the ventral vagal pathway that controls most of the key organ functions. As noted above, it encourages social engagement and interaction to further secure and stabilize the individual. The more recently recognized but older pathway, the dorsal vagal, controls the emergency freeze response, which causes immobility, lightheadedness, speechlessness, fainting and shock.

While the ventral vagal parasympathetic response is mediated by the neocortex, the newest and most developed part of the brain, the dorsal vagal parasympathetic response is mediated or activated by the most primitive, reptilian part of the brain.

Malfunctioning of either of these vagal pathways can lead to emotional disturbances, but regulating the vagal tone can moderate the disturbances. Brain function, specifically emotional responses and reactions, are directly affected by signals carried by the vagus nerve. Studies have shown that behavioral measures of emotional expression, emotional disturbances, self-regulatory skills, and reactivity may be correlated with baseline cardiovascular levels of vagal tone, leading to the conclusion that cardiovascular vagal tone can be an indicator of how well emotions are being regulated and managed. This perspective was not under consideration traditionally until the Polyvagal Theory opened this enlightened perspective and continues to encourage further experimentation.

The higher the level of vagal tone, the healthier the baseline condition of mind and body. Therefore, given the direct relationship between physical and emotional conditions, it follows that practicing the exercises to improve physical vagal tone will contribute to the improvement of emotional conditions, returning them to more normal baseline levels.

Emotional conditions that may be the consequence of low vagal tone include anxiety, depression, sensations of stress, fatigue not caused by excessive activity, and sleeplessness. Other, more long lasting emotional conditions may include Post Traumatic Stress Disorder (PTSD), and Attention Deficit Hyperactivity Disorder (ADHD). While many of these emotional disorders may respond to professional counseling and prescribed medication, hard-to-treat cases may respond favorably to vagal toning activities.

Physical actions that can return the body's emotional and physical reactions to normal baseline levels, as discussed in some of the previous chapters, include Yoga stretches and poses, various forms of meditation, oral exercises to stimulate the vagus nerve in proximity to the vocal cords, cold water to the face, auricular massaging of the ears and earlobes and sides of the neck to stimulate the vagus nerve as it passes through the ears and along the carotid arteries. Practicing mindfulness, or being in the moment, is a variation on meditation, with awareness of every environment stimulus.

The effectiveness of all of these exercises can be enhanced by managed, diaphragmatic breathing, with deep, deliberate, thoughtful inhales and exhales, which directly stimulate the vagus nerve. The effect is to slightly increase the heart rate on inhales, and them to lower heart rate back to a healthy, or homeostatic baseline on exhales.

When vagal tone is high, physical and emotional states are normal. Low vagal tone, the consequence of not stimulating the vagus nerve, can result in the range of emotional disorders we've been discussing and, additionally, can contribute to a sense of apathy, loneliness, isolation, and negative moods. These are all symptoms of the inability to engage socially and participate in social interaction. This may continue a self-perpetuating downward spiral, with the sense of isolation tending to discourage social interaction, and with the disconnection from social engagements furthering the feelings of isolation.

Low vagal tone can have equally serious consequences physically, including cardiovascular disorders.

Cardiovascular Applications

The relationship between the vagus nerve and the heart has been extensively researched and verified, with further clarification emerging from the Polyvagal Theory.

To set the stage for understanding this relationship, let's begin with the physical side of the relationship, keeping in mind that the vagus nerve is neither organ or muscle, but is a long, multi-branched network of wire-like nerves connecting the brain and other organs, carrying the electrical impulse messages between them. This is how the vagus nerve carries messages that affect one of the most important physical elements, the heart, and does so every second, 24 hours every day.

The vagus nerve travels from the brainstem and connects with the heart muscle or myocardium on the upper right side of the heart, in a cluster of nerves called the sinus node, for short, or sinoatrial node. Here the vagus nerve acts like a natural pacemaker, regulating the heartbeat. During normal conditions, at times of homeostasis, when there is little or no activity or stress, signals arriving from the brain through the vagus nerve slow the heart rate to less than 100 beats per minute. It is subsequently slowed and regulated, sequentially, by the atrioventricular node, the bundle of His, the right and left bundle branches, and finally the Purkinje fibers at the bottom of the myocardium. Every second or so, the heart muscle contracts, blood is forced out of the ventricles toward the lungs from the right ventricle, and into the aorta from the left ventricle.

Now, here is where the relationship between the heart and emotional reactions occur, but first, a quick background. The Polyvagal Theory has added clarity to our understanding of how the autonomic nervous system in primates evolved from the more primitive reptile nervous system. Changes evolved to accommodate the more complex primate nervous system, resulting in increasingly elaborate vagal pathways that control or regulate the heart. There was a transition from the exclusive dorsal vagus nucleus among reptiles to a more elaborate structure in mammals, called the nuclear ambiguus.

This included a connection between the heart and the face that enabled social interactions to influence the visceral or bodily functions, and possible dysfunctions. In simple terms, this means that social activity and other emotionally-regulated activities could play a role in maintaining control over the heart rate, while conversely, cardiovascular events can directly affect the emotions.

Charles Darwin, the founder of evolutionary theory, recognized the bi-directional flow between the brain and the heart that is mediated by the vagus nerve. Darwin understood that facial expressions were a physical manifestation of emotions, and correctly surmised that there were neural pathways connecting the brain with the heart and other organs that would facilitate physiological responses to emotions. Darwin and those of his time were correct in their estimate, despite not yet knowing that the *pneumogastric nerve*, later renamed the vagus nerve, had its own private network connection between the brain and the heart, apart from the connections of the action-oriented sympathetic nervous system. Capabilities to elevate and reduce heart rate coexist.

Today, Polyvagal Theory has led to discoveries of how vagal tone, the state of homeostasis, can be determined. A simple but effective determination of vagal tone is measurement of the heart rate during inhalation, when it should increase slightly above baseline, and then, measurement of the heart rate during exhalation, when the heart rate should return to baseline. The different rates of the two heart rates can be used to specify the precise vagal tone.

What does this mean to you?

During times of stress, your physical side may be in a state of elevated heart and respiratory rate, and you may be sweating, feeling a need to exert yourself and take action. In those situations when the cause of the sympathetic response is alleviated, and there is no need to run, or fight, or jump, you can bring things down, calm your body, with thoughts of calm, peace, reassurance. Repeat to yourself that everything is cool, under control, and it's okay to relax.

On a more serious medical level, when controlling heart rate and respiration are beyond the self-application of physical and emotional exercises, a relatively new treatment is the subcutaneous insertion of a vagus nerve stimulator. It is connected to the vagus nerve, and generates regular electrical impulses, acting like a pacemaker to increase vagal tone. Newer technologies have led to non-invasive electrical stimulators that make access to this treatment less expensive, and enables application to a wide range of symptoms. One of the first successful applications of electrical vagus nerve stimulation has been treatment of epilepsy, and up to

50% success in reduction in seizures has been achieved, among patients not responding to medication.

Autoimmune Responses and Inflammation

A relationship has been established between the autonomic nervous system (ANS) and the body's inflammatory response. It has long been understood that the autoimmune system includes inflammation among its responses to infection, since inflammation helps trigger many aspects of the body's defense, including release of macrophages or white blood cells, and killer T-cells that identify and annihilate invading microorganisms. But often the autoimmune system can overreact and overwork, continuing inflammation to the point that it can become damaging.

Non-drug treatments to calm the autoimmune responses are being derived from Polyvagal Theory. One approach is rocking, that is, a rocking motion in a chair or on a cushion. This is believed to have a soothing effect overall, and a stimulating effect on carotid baroreceptors. Recall that vagal tone can be increased by massaging the vagus nerve on both sides of the neck, where the vagus nerve runs past the carotid arteries. As a result of steady, continuous rocking several times a day for several days, blood pressure levels are lowered as the relaxation functions of the parasympathetic nervous system are engaged.

Another relaxant of the autoimmune response involves contractions of the pelvic floor, in a manner similar to contractions of the diaphragm. But while the diaphragm controls the upper body functions of the lungs and respiratory system, the pelvic floor affects the lower body, including the bladder and colon. An exercise to contract and engage the pelvic floor involves sitting on an exercise ball and feeling the pelvic floor begin to relax and settle into the ball, then trying to tighten it, then releasing it, letting it settle again, and repeating the cycle.

Dr. Stephan Porges, founder of Polyvagal Theory, also advocates standing on a half exercise ball with a rounded bottom and flat top, with someone else holding the person's hand to steady and give reassurance. This not only facilitates the therapeutic benefits of the balancing effort, it introduces a social engagement function, which signals the calming parasympathetic nervous system to initiate the socially engaging and relaxing ventral vagal response.

Added to these targeted, specific exercises can be the group of actions that have been used for other situations where the fight or flight sympathetic nervous system has engaged and needs to be turned down, or whenever the dorsal vagal response creates immobility, lightheadedness and more severe freeze symptoms. These include Yoga poses and stretches, meditation, vigorous cardiovascular exercises, massage of the neck and ears, cold facial therapy, and importantly, diaphragmatic deep, conscious breathing.

Autoimmune reactions as discussed here are moderate and are not at the level of being serious, chronic or life-threatening. But in cases of more serious autoimmune disorders, there is no substitute for professional medical treatment. The critical first step is the correct diagnosis of the condition and identification of its cause.

Our contemporary ingestion of medications for numerous conditions, both real and imagined, can lead to bodily reactions, notably autoimmune overreactions. This may be exacerbated by taking herbal supplements, which can conflict with medications being taken, or that might initiate autoimmune disorders on their own:

> Herbal supplements are lightly regulated by the Food and Drug Administration (FDA), and

marketers may not be fully cognizant of potential side effects. Anyone taking prescription medications should check with their physician or pharmacist before mixing their medications with herbs.

Chapter 9: Clinical Applications of Polyvagal Theory

Facial Expressions, Asperger's Spectrum, and Autism

The Polyvagal Theory addresses the treatment of autistic and other, less extremely affected Asperger's Spectrum children, with the presupposition that these children's social interaction capabilities are physically undamaged, and may be awakened with the right type of stimulation. Given that many of these afflicted children are unable to control their social behaviors, or more precisely, unable or unwilling to activate and use their social behavior voluntarily, the Polyvagal Theory holds that there are ways to stimulate the vagus nerve in ways that can encourage the children to manifest the physical dimensions of social engagement.

The Polyvagal model assumes that for many Asperger's children with social communication deficits, including those at the extreme end who are diagnosed with autism, their social engagement systems are intact and they are not missing or irredeemably damaged components of their central nervous systems.

In recalling the earlier discussion of *neuroception*, this is a concept with the purpose of analyzing and interpreting certain environmental factors, and then to initiate either defensive reactions, or to stimulate the onset of the calming reaction. An example is the function of neural circuits that enable a child to smile and respond positively when recognizing someone familiar, but to hesitate or flee from an unknown stranger. These reactions are common among all children and can be overcome by simple reassurances.

But in situations involving Autism, and less extreme Asperger's Spectrum disorders, the goal is to find ways to arouse, or initiate, the positive responses to the familiar, and suppress the escape or avoidance tendencies that are almost always functioning. The autistic and Asperger's children, according to Polyvagal Theory, are in a permanent state of fear-inducing unfamiliarity and need to be drawn out.

It has been found that autistic and Asperger's children may be engaged socially by the use of encouraging, reassuring facial expressions, altering neuroception. In test situations, it is found that the social inhibition of children with autism may be less of a physical disorder than a purely emotional reaction to stress. If this is correct, it may be theoretically possible to associate their symptoms with either low level sympathetic defensive responses to stress and fear, or possibly due to dorsal vagal freeze immobilization. The combination of facial expressions, especially wide, sincere smiling and eyes wide open, eye-contact and reassuring speaking can begin to give autistic children a sense that they can trust someone, and begin to socially interact with the person. This is consistent with other studies that have demonstrated that social engagement can contribute to the calming, relaxing parasympathetic response.

In contrast, however, new studies are finding that certain brain anomalies can inhibit facial recognition in autistic children and teenagers. In these instances, there is a physical barrier that increases autism symptoms and reduces the potential for facial expression therapy to be effective.

Vagus Nerve Dietary and Nutritional Influences

Among the newer discoveries that involve the vagus nerve, studies are drawing a vagal connection between what we eat and how the brain responds.

New research into obesity control and the role of various types of diets and foods has identified a unique and important role of the vagus nerve in transmitting data from the stomach to the brain. The vagus nerve connects to nerves in the stomach, and sends that afferent data to inform the brain of the caloric value, or potential energy, of the stomach's contents. These data, in turn, causes the brain to either suppress appetite-stimulating hormones, when calorie counts are high, or to increase these hormones when the calorie content is low.

Controlled studies have been conducted among volunteers who agreed to hospital confinement so their exact consumption behavior can be accurately recorded. The usual method of having study participants record their eating experiences in a diary has been found to be highly inaccurate.

The researchers have discovered that the data forwarded to the brain by the vagus nerve can be distorted by over-processed foods and especially by artificial sweeteners. In the case of saccharine-type sugar substitutes, the part of the brain responsible for decision-making, the striatum, is misinformed, interpreting the afferent information to mean there is a specific energy potential available in the gut. When the expected energy is not available, the brain actually becomes confused, and encourages more eating, leading to ingestion of excessive calories.

In an environment of many, often conflicting, dietary recommendations and claims, each purporting to be ideal diets for health and for weight control, these new findings strongly discourage diets overly based on over-processed foods, and dietary consumption of artificial sweeteners. Natural, unprocessed, whole foods, long the standard of our evolutionary ancestors, remain the more responsible nutritional choice.

Electrical Vagus Nerve Stimulation

To expand upon the previous discussion of electrical vagus nerve stimulation, this will provide further insights into the methodology and results of the therapy to date. Electrical impulse therapy, called vagus nerve stimulation (VNS) may be prescribed when a patient with certain disorders, including epilepsy and depression do not respond well to traditional treatments and medications.

VNS is a form of neuromodulation. The concept is similar to a pacemaker that provides electrical impulses to the heart to ensure a continuous, regular heartbeat. For vagus nerve therapy, a small electrical generating device is surgically inserted subcutaneously in the chest, and a thin wire from the device is wrapped around the left vagus nerve in the neck, near the left carotid artery cavity. Branches of the vagus nerve travel down the left and right sides of the neck and chest, but the right side nerve is considered too risky because it has branches that supply connective fibers to the heart.

When activated, the device emits mild electrical currents that enter the vagus nerve and travel to the brainstem where the vagus nerve and other cranial nerves originate, and then on to parts of the brain. The electrical device is programmed to send impulses on a regular basis, automatically, generally with no voluntary action required. However, in cases when an epileptic seizure is beginning, a magnet can be passed over the generating device to increase the magnitude of the electrical impulse. This works in some cases to stop a seizure before it can intensify.

The exact way the VNS works to control seizures is not fully understood. It is believed to increase the blood supply to areas of the brain where seizure-inducing impulses occur. Another possible action is the increase of neurotransmitters, which are the chemicals that react between nerve dendrites, or end branches, to facilitate passage of electrical impulses between adjacent nerves. Another function is based on electroencephalogram (EEC) studies that show increased heart rate immediately before the onset of a seizure. Accordingly, some of the newer devices can respond to sudden increases in heart rate by sending an electrical surge to stop the seizure.

VNS therapy is especially useful for people with refractory epilepsy, which is drug-resistant and two or more medications do not stop seizures. The effectiveness of VNS therapy varies, with increasing effectiveness over time. Seizure are averagely reduced by 36% during the first three to six months, 58% after four years and 75% after 10 years. Users of the device report improved quality of life and reduced severity of seizures when they do occur.

In another study among epilepsy patients, a test was conducted applying either low level or high level electrical impulses. The 94 patients receiving the higher level impulses averaged a 28% reduction in seizures, compared to a 15% reduction for the sample of 104 patients receiving the lower level impulses.

As an alternative to the surgically implanted devices, new non-invasive electrical stimulation devices have been introduced. Initial approval has been granted in Europe for treatment of epilepsy, depression and pain, while in the U.S., the Food and Drug Administration (FDA) approval is currently limited to treatment of cluster headaches.

While vagus nerve therapy for epilepsy is given concurrent with medications, there is optimism that over time, the need for medications will be reduced as the effectiveness of electrical stimulation in improved.

Conclusion

The Polyvagal Theory has broadened medicine's understanding of our autonomic nervous system's origins and functions, opening up new therapies for treating a wide range of disorders, from anxiety and depression to epilepsy, Asperger's Spectrum, including autism. Knowing the full extent of the vagus nerve's reach and influences has made these treatments and therapies possible.

Of immediate, practical value on a personal, ongoing basis are the Polyvagal applications that can be easily learned and practiced. Even those with no debilitating or life-affecting issues can benefit from these practices, which can lead to a more consistent state of calm and peacefulness, and a positive physical and mental state. Evidence suggests that overcoming and managing stress can reduce inflammation and help protect against the onset of serious physical and emotional disorders.

Polyvagal Applications You Can Practice

The Polyvagal Theory, since its inception in 1994, has established expanded knowledge of the relationships between our physiological responses and our emotional responses. The autonomic nervous system's sympathetic and parasympathetic nervous systems are now understood to have wide-ranging influence over us.

> Physical actions, as simple as facial expressions and body language, or more proactive, conscious, managed breathing exercises can invoke emotional reactions of consequence.

> Emotional reactions, to danger, for example, can incite strong physical responses, from adrenaline-driven fight or flight reactions to immobility and complete inaction.

The autonomic nervous system plays a larger role than expected. It is now understood that disorders including anxiety, depression, loneliness, and isolation may be the result of the low-level continuation of the sympathetic nervous system's action-orientation.

In more extreme cases, Asperger's Spectrum conditions, including autism, may be caused by these overreactions of the defensive sympathetic nervous system and can be treated by stimulating the vagus nerve to activate the parasympathetic nervous system's calming, relaxing social engagement functions. These treatments leverage physical and emotional vagal exercises employing facial expressions, eye contact and body language to reduce the uncertainties and inability to engage with others.

And while the counterforce of the stabilizing, calming parasympathetic response may bring down the sympathetic-induced faster heart rate, and faster, more shallow breathing, it may not do so fully and needs our conscious help and actions to return our bodies and our nervous systems to stability and homeostasis. This help can be our application of vagal toning exercises, including massage to the neck and ears, Yoga poses and stretching, meditation and mindfulness, cold therapy to the face and body, and especially the rapid vagal toning achieved by deep, deliberate, conscious breathing, accompanied by diaphragmatic expansions and contractions.

Polyvagal Theory has also raised awareness of the parasympathetic nervous system having not one but two dimensions, which are the well-known socially engaging, calming response, now called the ventral vagal response, and the newly recognized primal response of freezing in place, immobilized. This is called the dorsal vagal response. These two opposing vagal responses can bring us to normalcy, or can shut our autonomic nervous system down totally, potentially causing us to faint or go into shock.

> For clarity, the term opposing is applicable in comparing the ventral and dorsal vagal responses, since the former creates normalcy or homeostasis, and the latter creates immobility and disablement. In reality, the dorsal freeze appears to be an extreme ventral-type response in that it greatly exaggerates calming.

Medical Applications

Self-management of autonomic nervous system emotional and physical trauma can extend to a range of possibilities, as just discussed, but individuals whose condition is more serious can benefit from professional assistance.

As noted in a previous chapter, chiropractic doctors working with Asperger's patients have come to recognize that Polyvagal Theory is compatible with their *salutogenic* healthcare model, by supporting the recognition that the body can self-regulate itself, and it can self-heal under the right conditions. This inspired recognition that the Polyvagal response can enable the doctors to tap into its healing potential. They apply neurological exercises that stimulate vagal tone in their patients, and this provides the Asperger's Spectrum patients with new ways to hear, to perceive, to respond to people and situations, to smile, to speak, to maintain eye contact.

Further along on the medical level, electric stimulation of the vagus nerve is successfully being applied to treat epilepsy, notably in cases where the patient has not responded well to drugs. A small electrical device is implanted subcutaneously, and a thin wire extends to connect with the vagus nerve. When spasm-inducing impulses are detected by the device, it emits an electrical pulse that stimulates the vagus nerve to send a signal to the brainstem, which in turn transmits the impulse to the part of the brain controlling involuntary seizures. To-date, results are successful in up to 50% of cases that have otherwise not responded well to treatment with multiple medications.

Electrical stimulation is also being applied in cases of extreme depression, not controlled by medication or electroshock therapy. Results in reducing depression have been mixed and experimentation continues. Other conditions responding, to some degree to electrical vagal stimulation include Parkinson's Disease, cluster headaches, rheumatoid arthritis and irritable bowel disease.

Recently, non-invasive electrical vagus nerve stimulation devices that do not require surgical implantation have been developed, promising a wider range of possible applications. European authorities have authorized a range of applications, but so far, in the U.S., testing has been mostly limited to treating cluster headaches.

Non-Medical Applications

Polyvagal Theory explorations into social engagement as a fundamental aspect of the parasympathetic nervous system have opened the social cue to engagement, notably facial expressions and body language. In effect, people are more affected emotionally and physically be the smiles, degrees of eye opening, tones of voice and other bodily gestures than previously understood. Apart from medical applications, such as for Asperger's Spectrum disorders, there are commercial and other non-medical applications that may be on the near horizon.

Body language has been considered a signal of subconscious motivations and intentions since it was publicized decades ago. For example, a person you are speaking with whose arms are crossed is actually signaling either disagreement or rejection of what is being said. A person who will not make eye contact with you may not be sincere and may not be trusted, and a person whose hand covers their mouth, or whose hand is held close their mouth while speaking to you may be lying. Someone who smiles sparingly, rather than widely, is less than genuinely interested in you or in what you are saying. If someone turns away from you, or leans away, they may be demonstrating a lack of respect to you.

The advice columns and popular books that presented these concepts to the public offered not only the caution to be aware of these physical cues, but to be equally careful not to unconsciously tip off or signal the other person with your own body language and facial expressions.

Today, subconscious signals are being elevated to a higher, more scientific level. Thanks to facial recognition software, a person's facial expressions can be almost instantly analyzed to a very precise degree, far more sophisticated than a person can determine themselves.

Facial recognition software is now being enhanced by artificial intelligence so that its precision in accurately assessing subconscious thinking is unsurpassed.

One example of the application of this enhanced technology is in meetings and negotiations. Imagine this scenario:

> You are an executive, attending a meeting at your company headquarters with other executives from a company that is inviting an investment from the company you represent. You listen carefully to what is being said, and what is written on the documents in front of you.
>
> But at the same time, a hidden camera is tracking the facial expressions of the visiting executives, who are requesting the investment. While things appear to be progressing positively, and the proposed deal looks attractive, you have another, valuable input in front of you. On your mobile phone screen, a rating of honesty and sincerity is flashing numerical scores across a dozen criteria, as the hidden camera has been tracking the facial expressions and body movements and has determined that something is not right.

Cautions are flashing on your screen. While your technical team will later provide an in-depth analysis, you already know not to agree to the terms, and request some time to consider. Later, the analysis informs you that the visiting executives have been deceptive in what they have said and further appear to be withholding important information. A lawsuit, perhaps? Or a court injunction, a regulatory agency action, or a competitive threat?

Other applications of this facial recognition and body language analytical software can extend to advertising. Imagine a small electronic billboard in a shopping mall. As a shopper approaches, the facial expression that a camera on the billboard immediately recognizes gender, age and ethnicity, and additionally assesses the state of mind of the shopper. If the shopper's facial expression shows skepticism, the advertisement can adapt the face of the spokesperson to increase trust and interest.

Reference List

Porges, S. (2001) The Polyvagal Theory: Phylogenetic substrates of a social nervous system
https://www.sciencedirect.com/science/article/abs/pii/S0167876001001623

Porges, S. (2009), The Polyvagal Theory: New insights into adaptive reactions of the autonomic nervous system, *Cleveland Clinic.*
https://www.ncbi.nlm.nih.gov/pmc/articles/PMC3108032/

Polyvagal Theory, (2019), *Wikipedia*
https://en.wikipedia.org/wiki/Polyvagal_theory

Small, D., Shell, E., (October, 2019) Obesity On The Brain, *Scientific American,* Vol 324, No. 1.

Rosenthal, M. (2019), The Science Behind PTSD Symptoms: How trauma changes the Brain
https://psychcentral.com/blog/the-science-behind-ptsd-symptoms-how-trauma-changes-the-brain/

Puder, D., (2018), Emotional Shutdown—Understanding Polyvagal Theory, *Psychiatry & Psychotherapy* Podcast.
https://psychcentral.com/blog/the-science-behind-ptsd-symptoms-how-trauma-changes-the-brain/

Burkett, D. (2009) Bradyarrhythmais and Conduction Abnormalities, *Small Animal Care Medicine*
https://www.sciencedirect.com/topics/veterinary-science-and-veterinary-medicine/vagal-tone

Porges, S. (2018), Clinical Applications Of The Polyvagal Theory, *W.W. Norton*

National Institutes Of Health, (2019) What are the parts of the nervous system?
https://www.nichd.nih.gov/health/topics/neuro/conditioninfo/parts

MedlinePlus. (2016). *Neurosciences*
https://medlineplus.gov/ency/article/007456.htm

Newman, T. (2017), All about the central nervous system. *Medical News Today*.

https://www.medicalnewstoday.com/articles/307076.php

The Autonomic Nervous System (2019), *University of Washington*,
https://faculty.washington.edu/chudler/auto.html

Samsel, M. (2019) The Dorsal Vagal Shift,
https://reichandlowentherapy.org/Content/Vegetative/dorsal_shift.html

Samsel, M. (2019) Shift Toward The Social Engagement System,
https://reichandlowentherapy.org/Content/Vegetative/ventral_shift.html

Polyvagal Theory Revisited, (2019), *Therapist Uncensored*,
https://www.therapistuncensored.com/increase-your-cool-managing-your-ventral-vagal-system/

Basic Model of Social Communication (2019), *Management Media*,
https://managementmania.com/en/basic-model-of-social-communication

Porges, S. (2019), Podcast: Face-To-Face Interactions & The Social Engagement System, *ILS Integrated Listening Systems*, https://integratedlistening.com/ils-podcast-series-library/

Wikipedia, (2019), Social Engagement Definitions, Characteristics, https://en.wikipedia.org/wiki/Social_engagement

Porges, S. (2014) Social Engagement Heals, https://attachmentdisorderhealing.com/porges-polyvagal3/

Bunn, T., (2014) Using The Social Engagement System, *Psychology Today*, https://www.psychologytoday.com/us/blog/conquer-fear-flying/201211/using-the-social-engagement-system

Saraswati, S. (2019) How to use relaxation to reduces the effects of stress, handle traumatic events...feel safe and secure, https://www.bigshakti.com/how-to-use-relaxation-to-reduce-the-negative-effects-of-stress-handle-traumatic-events-prevent-overwhelm-and-feel-safe-and-secure/

Garner, G. (2019), Nervous System Hacks To Keep Calm and Vagus On, https://gingergarner.com/nervous-system-hacks-to-keep-calm-vagus-on/

Porges, S. (2019) How to Manage Anxiety With a Breathing Exercise *School of Modern Psychology*, https://www.schoolofmodernpsychology.com/blog/manage-anxiety-with-a-breathing-exercise-by-dr-stephen-porges

Neural mechanisms of mindfulness and meditation: Evidence from neuroimaging studies, (2014), *World Journal of Radiology*, https://www.ncbi.nlm.nih.gov/pmc/articles/PMC4109098/

Yoga Therapy and Polyvagal Theory: The Convergence of Traditional Wisdom and Contemporary Neuroscience for Self-Regulation and Resilience (2019). https://www.ncbi.nlm.nih.gov/pmc/articles/PMC5835127/

Ehrmann, W., (2019), The healing role of breathwork for depression based on the Polyvagal Theory, https://www.wilfried-ehrmann.com/wp-content/uploads/Polyvagal-e.pdf

Porges, S, (2011) The Polyvagal Theory: Neurophysiological Foundations of Emotions, Attachment, Communication, and Self-Regulation. *W W Norton & Co.*

Rubin, D. (2019), Autism, Neuroplasticity, and the Polyvagal Theory…What Does It All Mean? *Pathways To Family Wellness*, https://pathwaystofamilywellness.org/Chiropractic/autism-neuropasticity-and-the-polyvagal-theory-what-does-it-all-mean.html

Bridges, H. (2017), Podcast: The remarkable benefits of the PVT for those on the spectrum, https://www.youtube.com/watch?v=t_Th_t7C8BM

Porges, S. (2019) Brain-body connection may ease autistic people's social problems, *Spectrum*, https://www.spectrumnews.org/opinion/viewpoint/brain-body-connection-may-ease-autistic-peoples-social-problems/

Bertone, H., (2019), What meditation is all about, *Healthline,* https://www.healthline.com/health/mental-health/types-of-meditation?slot_pos=article_2&utm_source=Sailthru%20Email&utm_medium=Email&utm_campaign=general_health_B&utm_content=2019-10-03&apid=25264436

Gaia Staff, (2016) Which type of meditation style is best for you? Gaia, https://www.gaia.com/article/which-type-meditation-style-best-for-you

Land, R. (2019) 10 Simple Yoga Poses, *Yoga Journal,* https://www.yogajournal.com/poses/10-simple-yoga-poses-that-help-everyone-at-any-age#gid=ci02228525500026fe&pid=05-tree

Wang, Y. (2009), Ear Acupuncture, Micro-Acupuncture in Practice, https://www.sciencedirect.com/topics/medicine-and-dentistry/auricular-acupuncture

Roizen, M. (2019), How does acupuncture affect the vagus nerve?

https://www.sharecare.com/health/acupuncture/acupuncture-vagus-nerve

OSEA, (2018), Massage for Vagus Nerve Stimulation, https://oseamalibu.com/blogs/wellness-blog/massage-for-vagus-nerve-stimulation

National Institutes of Health (NIH), (2019), Auricular Acupuncture and Vagal Regulation, https://www.ncbi.nlm.nih.gov/pmc/articles/PMC3523683/

Bergland, C., (2017), Diaphragmatic Exercises and Your Vagus Nerve, *Psychology Today,* https://www.psychologytoday.com/us/blog/the-athletes-way/201705/diaphragmatic-breathing-exercises-and-your-vagus-nerve

Pasquier, M., Clair, M., et al, (2019), Carotid Sinus Massage, *New England Journal of Medicine,* https://www.nejm.org/doi/full/10.1056/NEJMvcm1313338

Johns Hopkins Medicine, (2019), *Overview of Nervous System Disorders,* https://www.hopkinsmedicine.org/health/conditions-and-diseases/overview-of-nervous-system-disorders

Central Nervous System Fatigue, (2019), *Wikipedia*, https://en.wikipedia.org/wiki/Central_nervous_system_fatigue

Miller., L. (2016), The Stress Solution, The Different Kinds of Stress, *American Psychological Association*, https://www.apa.org/helpcenter/stress-kinds

The link between ADHD and anxiety, (2019), *Healthline*, https://www.healthline.com/health/adhd-and-anxiety

Cozolino, L., (2019), The Tribune Brain: Three Brains Trying to Work as One, *Psychotherapy Network*, https://www.psychotherapynetworker.org/blog/details/48/the-triune-brain-three-brains-attempting-to-work-as-one

McGill University, (2019), The Evolutionary Layers of the Human brain, https://thebrain.mcgill.ca/flash/d/d_05/d_05_cr/d_05_cr_her/d_05_cr_her.html

National Institutes of Health, (201910, Cognitive Impairment and Rehabilitation Strategies After Traumatic Brain Injury, https://www.ncbi.nlm.nih.gov/pmc/articles/PMC4904751/

PTSD: Post Traumatic Stress Disorder, (2019), The Mayo Clinic, https://www.mayoclinic.org/diseases-conditions/post-traumatic-stress-disorder/symptoms-causes/syc-20355967

Schwartz, A., (2019), The Polyvagal Theory and Healing Complex PTSD, https://drarielleschwartz.com/the-polyvagal-theory-and-healing-complex-ptsd-dr-arielle-schwartz/#.XZt1XCV7k_U

Wagner, D., (2019), Polyvagal Theory in Practice, *Counseling Today*, https://ct.counseling.org/2016/06/polyvagal-theory-practice/

Bradbury, T. (2017), 8 Ways to Read Somebody's Body Language, *Inc.com*, https://www.inc.com/travisbradberry/8-great-tricks-for-reading-peoples-body-language.html

Crowley, OV, et al, (2011), The interactive effect of change in perceived stress and trait anxiety on vagal recovery from cognitive challenge, *Columbia University Medical Center, Division of Behavioral Medicine,* https://www.ncbi.nlm.nih.gov/pubmed/21945037

Beauchaine, T. (2001), Vagal tone, development, and Gray's motivational theory: Toward an integrated model of autonomic nervous system functioning in psychopathology, Cambridge University Press, http://tpb.psy.ohio-state.edu/papers/VT.pdf

Rigoni, T., (2017), The Role of Baseline Vagal Tone in Dealing with a Stressor during Face to Face and Computer-Based Social Interactions, *Frontiers in Psychology,* https://www.frontiersin.org/articles/10.3389/fpsyg.2017.01986/full

Porge, S., Kolacz, J., (2018), Neurocardiology Through the Lens of the Polyvagal Theory, *Indiana University Department of Psychiatry,* https://static1.squarespace.com/static/5c1d025fb27e390a78569537/t/5cb67958eef1a19d9940fb5d/1555462489535/Neurocardiology+final+Porges+Kolacz.pdf

Porges, S. (2004), Neuroception: A Subconscious System for Detecting Threats and Safety, Institute of Education Services,
https://eric.ed.gov/?id=EJ938225

Porfes, S., et al, (1994), Vagal tone and the physiological regulation of emotion. *Institute for Child Study, University of Maryland.*
https://www.ncbi.nlm.nih.gov/pubmed/7984159

Hitzmann, S., (2016), Why vagal tone is so important,
https://www.meltmethod.com/blog/vagus-nerve/'

Bergland, S., The neurobiology of grace under pressure, *Psychology Today,*
https://www.psychologytoday.com/us/blog/the-athletes-way/201302/the-neurobiology-grace-under-pressure

Trauma, Intimacy and the Polyvagal Theory, *Life Change Health Institute,* http://www.lifechangehealthinstitute.ie/discovering-importance-poly-vagal-theory/

Phillips, M. (2015), Healing Trauma and Pain Through Polyvagal Science, https://maggiephillipsphd.com/Polyvagal/EBookHealingTraumaPainThroughPolyvagalScience.pdf

CPSIA information can be obtained
at www.ICGtesting.com
Printed in the USA
BVHW041716121021
618766BV00011B/183